Carbon Capitalism

Carbon Capitalism

Energy, Social Reproduction and World Order

Tim Di Muzio

ROWMAN &
LITTLEFIELD
INTERNATIONAL
London • New York

Published by Rowman & Littlefield International, Ltd.
Unit A, Whitacre Mews, 26-34 Stannary Street, London SE11 4AB
www.rowmaninternational.com

Rowman & Littlefield International, Ltd. is an affiliate of
Rowman & Littlefield
4501 Forbes Boulevard, Suite 200, Lanham, Maryland 20706, USA
With additional offices in Boulder, New York, Toronto (Canada), and Plymouth (UK)
www.rowman.com

British Library Cataloguing in Publication Data

A catalogue record for this book is available from the British Library
ISBN: HB 978-1-78348-097-5
PB 978-1-78348-099-9

Library of Congress Cataloging-in-Publication Data

Di Muzio, Tim
Carbon capitalism : energy, social reproduction and world order / Tim Di Muzio.
pages cm
Includes bibliographical references and index.
ISBN 978-1-78348-097-5 (cloth : alk. paper) — ISBN 978-1-78348-098-2 (pbk. : alk. paper) — ISBN 978-1-78348-099-9 (ebook) 1. Energy security—Political aspects. 2. Energy consumption—Political aspects. 3. Fossil fuels—Political aspects. 4. Natural resources—Political aspects. 5. Capitalism. 6. Geopolitics. I. Title.
HD9502.A2.D525 2015
333.79—dc23

2015019532

∞™ The paper used in this publication meets the minimum requirements of American National Standard for Information Sciences—Permanence of Paper for Printed Library Materials, ANSI/NISO Z39.48-1992.

Printed in the United States of America

Rakkaalleni, Hannalle

Contents

Preface

It is likely that every book ever written never comes out exactly as one intended it at the stage of conception. Every study is 'a particular bundle of silences'—so Trouillot tells us—and this study is no different (1995, 27). It is a work set in a particular time period and written in particular places. There are always gaps to fill, time constraints and more literature to consult and critically reflect on, particularly the more we try to engage literatures outside our own narrow fields of specialization. But in the end, the author cannot escape making decisions of what to include and exclude, what to emphasize and what to downplay, how to weigh agency and structure and how to balance theory and history. The author, as it were, is a bit like a curator of words, thoughts, histories and theories. If Foucault is correct to say that knowledge is not made for understanding but for cutting, much has been left on the cutting room floor in the curation of this work (1984, 88). In truth, dear reader, this book could have been twice the length and filled with even more examples, but my partner reminds me that there will be other books to write. As you will come to find, what I have tried to do is provide a brief genealogy of what I call carbon capitalism and its concomitant petro-market civilization. I have done so, and the reader will have to judge how well, by focusing on the exploitation of fossil fuels, the forms of social reproduction that were made possible and the logic of differential capitalization. Readers who are interested in the history of capitalism will find a new theory of its emergence viewed from the capital-as-power perspective, one that takes energy, social reproduction and capitalization as fundamental aspects of the making of a world order. I hope the reader finds my arguments unique, well defended and insightful, but I will always welcome criticism. In this sense, I am well aware that this is not the final word on the subject. I have many people to thank, and the usual suspects know who they are. But this book would not have been possible without Hanna's patience and Matthew Dow's kind assistance with research. I also thank Tim Onslow for diligently preparing the index and Anna Reeve for supporting the project. I owe a great deal of this work to them. To conclude this short preface, I would like to leave you with a quote from Paul Sweezy that expresses my sentiments about this work almost exactly: "The conclusions so reached are obviously tentative; they are set forth here, however, because it seems likely that the insight gained by pursuing this method is sufficiently enlightening to warrant a great deal of further study along the same general line" (cited in Fine 1988, 237).

1

Carbon Capitalism and Petro-Market Civilization

In 2005 and 2007 a war game was played out by former high-ranking government officials in Washington DC. Entitled 'Oil Shockwave', the goal of the simulation was to analyze how policymakers might respond to a sudden oil shock brought on by conflict and political instability in Venezuela, Iran and the Caspian Basin. In the crisis scenario, a small amount of oil—4 percent—was taken off the world market, leading to a giant spike in oil prices from US$58 dollars a barrel to US$161. War in the Middle East, the return of the draft, petrol rationing at gas stations, extreme price inflation, increasing unemployment levels and the depletion of the US Strategic Petroleum Reserve (SPR) were all countenanced as plausible results of the crisis. Robert M. Gates, the Oil Shockwave national security advisor, noted that 'the real lesson here [is that] it only requires a relatively small amount of oil to be taken out of the system to have huge economic and security implications'.[1] The simulations were sponsored by Securing America's Future Energy (SAFE) and the National Commission on Energy Policy (NCEP). They were intended to highlight the United States' dependence on oil for its economic well-being and security, as well as to point out how years of policy inaction on renewable energy and oil dependence could lead to foreseeable, but potentially avoidable, catastrophes at multiple levels of the socioeconomic order of the United States—and by extension, most of the petroleum-soaked world.

What seems so strange about these simulations is that they followed two decades of cheap and abundant fossil fuels, uneven but generally growing economies, a revolution in telecommunications, the dissolution of the Soviet Union and the widespread belief in the globalization of capitalist markets and neoliberal development.[2] Of course, actual capitalism has never tended toward any equilibrium—real or imagined—but what was noticed in the simulations went beyond mere commonsense notions that capitalist markets could be unpredictable, unstable or prone to manias, corporate fraud and abuse (Kindleberger and Aliber 2005). The simulated crisis was not about the dangers of globalized markets and greedy irrational actors pursuing their own gain at the expense of society. Nor was it a crisis where the right mix of fiscal or monetary policy could come swiftly to the rescue to restore economic growth and employment. The simulated crisis was far more profound than the collapse of markets or the dangers of debt and financial alchemy; the crisis was so foundational that its scale and scope were civilizational.

Our leaders could be forgiven for their inaction and short-sightedness if this were the first time oil was recognized as essential to the social reproduction of a more globalized and capitalized world order. But they cannot. The fact that fossil fuels are nonrenewable sources of energy and that their production will eventually decline and become more expensive has long been discussed even before Hubbert's graphs on peak oil production in the United States were drafted.[3] Though there was prior recognition—particularly among the US and UK armed forces and some geologists—the oil price spikes of the 1970s were watershed moments in alerting entire populations to the extent their lifestyles had become reliant on stored sunshine (Crosby 2006, 62).[4] From 1972 to the early 1980s, when oil prices finally started to slowly decline, the price of oil—never its full ecological costs—had increased by an overall figure of 504 percent.[5] The price shocks helped generate double-digit inflation and skyrocketing unemployment in advanced economies and ballooned balance-of-payments deficits for most oil-importing countries, triggering an international debt crisis in the 1980s and the tombstone of the New International Economic Order (George 1988; Murphy 1984). The oil shocks also motivated the creation of the International Energy Agency (IEA) as a counter to the Organization for Petroleum Exporting Countries (OPEC). The latter was created in 1960 as an intergovernmental organization to coordinate the output and pricing of oil among major producers/exporters. The IEA, housed in Paris, was mandated to gather statistics, monitor the world energy situation and provide advice to its members on energy policy. Members of the organization—mostly net importers of oil—were also required to store a reserve of oil equivalent to ninety days of the previous year's imports.[6] The crisis even seeped into Anglophone popular culture in films like *Three Days of the Condor* (1974) and the *Mad Max* trilogy (1979–1985). The former, starring Robert Redford and directed by Sydney Pollack, was a fictional portrayal of how far clandestine security forces were willing to go to keep Middle Eastern oil flowing to the United States and its allies. The latter, starring Mel Gibson and directed by George Miller, fictionalized the breakdown of law and order in the Australian outback due to the scarcity of oil.[7]

At the same time, scholars of international relations were finding it difficult to ignore the importance of oil to international 'stability' and the global economy. Out of this concern and others emerged a new field of scholarly inquiry called international political economy (IPE). In the annals of the social sciences, IPE is a bit of a strange child. The subdiscipline is not the direct offspring of political economy (as one might suspect), but the outcome of considerable frustration with mainstream international relations (IR) theory (Di Muzio 2014, 161–62).

As many scholars have noticed, the field institutionalized and professionalized amidst the wrenching changes of the 1960s and 1970s.[8] Neocolonialism, the abandonment of the gold standard, floating exchange rates, rising oil prices, the limits to growth thesis, the environmental movement in the capitalist core, the debt crisis of the 'Third World' and the riddle of stagflation (among other events) all seemed to demonstrate that there were considerable flaws and omissions in mainstream accounts of international politics and power. Chief among them, of course, was the divorce of economics from politics (Dickins 2006; Hancock and Vivoda 2014; Philips 2005; Strange 1970; Underhill 2000). An additional concern, though not always taken up by self-professed international political economists, was the centrality of oil to the security and economic well-being of Western civilization. In the 1970s this could hardly be denied, and worries mounted whether the world energy crisis would precipitate the decline of American power and 'Western civilization' (Levy 1979; Lieber 1979). Some scholars in IR/IPE started to take energy seriously—particularly oil—but the vast majority tried to fit the importance of carbon energy into preexisting theoretical frameworks; used oil to 'test' their theories about international power, interdependence or both; or simply fretted about the power of OPEC and the short-term economic consequences of the price shock (Barnes 1972; Campbell 1977; Chubin 1976; Cleveland and Brittain 1975; Hallwood and Sinclair 1982; Hartshorn 1977; Healey 1979; Jabber 1978; Lewis 1974; Mikadashi 1980, 1981; Penrose 1979; Pollack 1974; Shackleton 1978; Smart 1977; Strange 1988; Turner 1976; Turner and Bedore 1978; Veit 1977; Willrich 1976). In this way, energy was made a fact of international relations and the global economy, but it never became a part of the *conceptual foundations* for understanding and explaining social relations, historical change and the constitution and reconstitution of world order over what Braudel (1983) called the *longue durée*.

There is only one book from that era that offered a foundational theorization of energy and human development reliant on critical political economy. A masterful work, now somewhat dated, Debeir, Deléage and Hémery's *In the Servitude of Power*, argued that the intimate link between capitalism and energy was vastly undertheorized in most critical analyses of human 'progress' and socionatural transformations (1986, xiii).[9] Regrettably, the book had the misfortune of arriving at a time when oil prices had dropped by 71 percent from their peak in 1980.[10] As a consequence, the book was virtually ignored by the literature in IR/IPE, not to mention elsewhere. Only recently has their work been rediscovered by a small coterie of scholars concerned with the role of energy in the history and social reproduction of capitalism and the future prospects of world order (e.g., Barca 2011; Huber 2008, 2013; Noreng 2007; Podobnik 2006; Spier 2011). Moreover, it is only within the last decade of increasing oil prices and the so-called 'War on Terror' that a litany of popular and scholarly studies on fossil fuels and the prospects of economic growth and renewable energy have appeared in droves.[11] This is perhaps not surprising given the massive 319 percent increase in the price of oil from 1999 to 2008, ruminations that the conventional supply of oil has already peaked, ongoing war, militarization and instability in the Middle East and arguments that our current patterns of fossil-fuelled–led development are ruining the planet's life-support systems for future generations and biodiversity (Friedrichs 2013; Hamilton 2004; Jackson 2009; Kempf 2008; Speth 2009).

This situation is largely attributable to the destruction and pollution of habitats and ecosystems, overconsumption by the 1 percent and the commercially affluent and the use of the atmosphere as a boundless sink for heat-trapping gases like carbon dioxide. However, with some exceptions, most of this recognition is happening outside of IPE and, as yet, seems to have had only a minor impact on the security studies literature (e.g., Colgan 2013a, 2013b; Elhefnawy 2008; Klare 2002, 2004, 2009). One telling illustration is the major textbooks or primers used to introduce undergraduate and graduate students to the study of IPE. Of the most renowned introductory texts in the discipline, one would be hard pressed to find any discussion of the importance of energy to the shaping and reshaping of world order and the global political economy, let alone the rise of 'hegemons' or 'great powers'. Where energy is mentioned at all in these texts, it plays a minor, rather than foundational, role in explaining historical change, the constitution of a more liberal world order and the transnationalization of production and exchange (Blyth 2009; Cohn 2011; Frieden and Lake 2003; Gilpin 2001; Hülsemeyer 2010; Miller 2008; Oatley 2011; O'Brien and Williams 2010; Palan 2000; Ravenhill 2008; Thompson 2000). So while these volumes are commendable in their own way, they are far from offering deep-seated theorizations on the importance of energy to social life and global order.

To my knowledge, there are only two main introductory texts on the global political economy that include a full chapter discussion on energy and international oil (Balaam and Dillman 2013; Gill and Law 1988).[12] What this suggests is that the study of the global political economy is largely disconnected from any material energy base and that IR/IPE students are likely to have a limited understanding of the importance of energy to human development and the formation and reformation of the global political economy, not to mention the political and livelihood consequences of high-energy modes of living.

RATIONALE FOR THE BOOK

Given the lack of critical attention paid by the field of political economy to the question of energy and its role in the constitution and reconstitution of world order in the IPE literature, this study purports to offer a historically informed, critical political look at what I will call carbon capitalism and a global rise and fall of the petro-market civilization. By 'critical', I mean at least three things: (1) an intellectual and moral stance whereby historical structures are not taken as self-evident but as institutions that have to be explained historically; (2) that we should question power relations as a matter of course and subject them to tests of legitimacy. If these chains of command and authority do not serve a legitimate democratic or socially desirable purpose, then they should be opposed, transformed and/or abolished; and (3) that where possible we should seek opportunities for greater freedom, emancipation and democratic practice in all forms of social organization (Bruff and Tepe 2011; Gill 2008).

Although it is true that IR/IPE is not fully blind to questions related to energy, as mentioned earlier, the literature largely focuses on the question of geopolitics, energy security and resource conflicts/wars, particularly as they relate to oil (Bromley 1991, 2005; Bunker and Ciccantell 2005; Colgan 2013a, 2013b; Elhefnawy 2008; Hiro 2006; Klare 2009; Labban 2008; Stokes 2007; Stokes and Raphael 2010). This research is incredibly important, but it has a tendency to be one dimensional and fails to conceive of the relationship between energy, the social reproduction of capitalism and world order on a civilizational scale from a critical political economy perspective. In this study, I characterize world order as a hierarchical petro-market civilization because of the radically unequal access to fossil fuel energy and centuries of Western domination, both violent and institutionally organized.

What I mean by the term *petro-market civilization* is an historical and contradictory pattern of civilizational order whose social reproduction is founded upon nonrenewable fossil fuels, mediated by the price mechanism of the market and dominated by the logic of differential accumulation (Di Muzio 2011, 2012; Gill 1995; Nitzan and Bichler 2009). Capital is the central institution of petro-market civilization and, as I will argue, its accumulation on a vast scale has been made possible by surplus fossil fuel energy while altering previous patterns of social reproduction tied more directly to photosynthesis and low-carbon energy growth (Goldstone 2002; Wrigley 2010). This is why I refer to carbon capitalism: the notion that the magnitude and universalization of capital accumulation, along with high energy–intensive forms of social reproduction, would have been impossible without abundant, affordable and accessible fossil fuels.[13]

This does not mean that other factors are unimportant; it is only to suggest the centrality of fossil fuels for understanding the present as history. The logic is relatively straightforward but nowhere articulated in the IPE literature in general and the literature on capital as power more specifically (Baines 2014; Brennan 2012; Di Muzio 2007, 2012; Hager 2013a, 2013b; McMahon 2013; Nitzan and Bichler 2009). The general argument can be stated in the following propositions:

- Energy can be defined as the capacity to do work; but what work will be done is a matter of differential social power enacted in physically violent ways, as well as through institutional forms of discipline and punishment.
- The uneven exploitation of coal, oil and natural gas at first provided small portions of humanity with surplus energy, mostly in England, Western Europe and the United States.
- Since energy is the capacity to do work, surplus energy means there is a greater capacity to produce, consume and exchange—albeit always conditioned by social property relations, the logic of differential accumulation and forms of political rule.
- Carbon capitalism can be conceived of as a conflictual mode of power where dominant owners control and capitalize the majority of this capacity to produce, consume and exchange for private benefit.
- This intraclass battle among investors/capitalists is recorded as differential capitalization measured in money. Here, capitalization means the present money value of owned income-generating assets.
- It is also an interclass mode of power whereby the 1 percent, or high-net-worth individuals, consume the majority of the planet's energy, accumulate it in monetary form and by their ownership of income-generating assets and their voracious consumption, have the largest ecological footprint (Di Muzio 2015a, 2015b; Kempf 2008).

However, these propositions cannot be understood in isolation or abstraction from social property relations, geopolitical competition for money and resources, a long history of Western colonialism, transatlantic slavery and violence, the constitution and reconstitution of gender orders and racialized labor hierarchies.

In sum, this book offers a new theorization of the global political economy that takes the energy basis of civilization, not as something we can tack on as auxiliary to political economy analysis, but as integral and inseparable for understanding and explaining its development, transformation and trajectories. This does not mean—and I stress here—that this book is about energy or resource determinism: the notion that energy source, supply, distribution, ownership, cost and use predetermine social formations and the precise fate of humanity. Rather, this study considers energy only within the context of human social property relations and the relation of force and struggle between rulers and ruled in the constitution and reconstitution of a world order I characterize as a hierarchical petro-market civilization. I will argue that the most important social relation of petro-market civilization is not between workers and capitalists per se, but between owners of income-generating assets and nonowners.

The second major rationale for the study is to apply perhaps the most convincing theorization of global capitalism to the study of energy, social reproduction and world order. This book is informed by the capital-as-power approach associated with the seminal work of Nitzan and Bichler (2009). The authors argue that capitalism should *not* be conceived of as a mode of production, but as a mode of power. In this framework, capital is theorized as *commodified differential power* measured in monetary units, and the focus is on the dominant firms and government organs at the center of pecuniary accumulation. This organized social power is commodified in capitalism because ownership titles or claims on this power can be bought and sold in the financial markets or privately transferred in law. Accumulation is understood as differential or relative, since investors capitalize a diverse set of income streams and benchmark their performance against rival accumulators, who likely hold a different portfolio of income-generating assets. What this means is that accumulation in this framework is understood as rising capitalization, such as growing bond yields and prices (depending on the position of the buyer/seller), increasing share prices and rising values for real estate and other investment products.

As we shall come to find out, oil and gas is the largest industry in the world in terms of capitalization when the estimated market value of state-run firms is taken into account (see Di Muzio 2012). If there is an epicenter to the entirety of modern international affairs, it is to be found in this sector of the global political economy—all else is secondary from the point of view of capitalization. But for Nitzan and Bichler, income or earnings are not a narrow offshoot of production, but the result of firms and capitalized governments actively shaping and reshaping the terrain of social reproduction by exerting energy over the entire social field.[14] Since earnings are primarily a matter of organized institutional power, then it is this power to create and re-create social reality that is essentially capitalized by investors. However, as I have argued, although Nitzan and Bichler have written extensively and insightfully, their work has largely underplayed the question of energy in the *historical development* of capitalism as a mode of power—though they by no means underplay oil and its importance to the global political economy (Di Muzio 2014).

By focusing on the transition to carbon energy consumption, I offer an entirely new account of capital's emergence and subsequent globalization.[15] I do so by theorizing three interlinked and equally weighted concepts: what I call the energy–capitalization–social reproduction nexus. The theorization of this nexus is explained in detail in chapter 2. In short, the second aim of this book is to provide an innovative, timely and novel study that mobilizes a new theory of capitalism to explain the rise and fall of carbon capitalism and petro-market civilization and reorient some of the key debates in the field of IR/IPE on hegemony, world order and global capitalism.

MAIN ARGUMENTS

This book makes a series of interconnected arguments that place at the forefront the importance of energy—particularly fossil fuels—to the global political economy of capital as power. Overall, these arguments may be considered a counterhistory to more dominant narratives that celebrate the 'rise of the West' and 'Western capitalism' as unproblematic, autogenerated and, on balance, of benefit to most of humanity—all the while ignoring the biospheric consequences of human actions (e.g., De Soto 2004; Diamond 1997; Ferguson 2011, 2012; Landes 1998). We will not be as extreme as Benjamin argues: "That all human history has been one giant catastrophe," (1968, 254) but nor will we ignore the power, domination and ecological degradation that served to constitute and reconstitute relations of force in both past and present.

Transitory Order Founded on Nonrenewable Carbon Energy and a New Periodization of Human History

The first argument is that our civilizational order can be conceptualized as an unequal and *transitory* petro-market civilization founded on fossil fuels as the dominant energy base for a considerable portion of humanity (Di Muzio in Gill 2011, 73–88). Since fossil fuels are nonrenewable deposits of stored solar energy, the civilizational order founded upon them must also be nonrenewable (cf. Engdahl 2012). What this means is that current patterns of social reproduction are historically unique in that they are decisively reliant on carbon energy, particularly oil. There are many interpretations of social reproduction, but what I mean by the term is the way in which any given society produces, consumes and reproduces its lives and lifestyles; how it conceptualizes these actions and how it defends them both discursively and materially—for example, in war or legal action. In saying this, I fully recognize the unevenness of carbon energy consumption across the world. However, given the prominence of international trade and geographically distant, yet integrated, supply chains—not to mention a petroleum-dependent global agribusiness system—a high dependence on fossil fuels in one place is likely to have direct or indirect consequences for lower energy economies.

The eruption of an Icelandic volcano on 15 April 2010 is but one example of the deep interconnections between the global north and south when it comes to carbon capitalism. The volcanic ash from the eruption grounded flights across Europe for days. Upwards of a million people travelling for business and pleasure were inconvenienced as airlines scrambled to adjust to daily losses of US$250 million and alternative methods of transport were found for some travellers.

But some green groups celebrated the grounding of most of Europe's air fleet. The inability to fly its kerosene-powered planes over much of Europe reduced European carbon emissions. Although one geologist estimated that the eruption of the Eyjafjallajokull volcano emitted 150,000 tons of carbon a day into the atmosphere, this pales in comparison to the daily dose of climate change–inducing gases emitted by the airline industry across Europe. According to the European Environmental Agency, normal flight traffic across thirty-two European countries emits 510,000 tons of carbon into the atmosphere every single day of the calendar year.[16] But the grounding of planes also revealed some of the international linkages that support the social reproduction of Europe's own petro-market civilization. Without a fully developed internal market of its own to absorb the products of fossil fuel–dependent industrial agriculture, scores of food bound for Europe on refrigerated cargo planes rotted in Kenya as the ash cloud hovered over Europe. According to the chief executive officer (CEO) of Fresh Produce Exporters Association of Kenya:

> Two million pounds of fresh produce is normally shipped out of Kenya every night. Eighty-two percent of that goes to Europe, and more than a third goes solely to Britain, whose airports have been among those shut down by the volcano's eruption. Five thousand Kenyan field hands have been laid off in the past few days, and others may be jobless soon. The only way to

> alleviate this would be to restore the air bridge to Europe, which would necessitate the equivalent of 10 Boeing 747s of cargo space—per night.[17]

In other words, not only is Kenya's capitalist-directed agriculture and the livelihood of its farm workers dependent on markets thousands of fossil fuel–propelled air miles away, but Europe's own diet largely relies on the delivery of foreign produce soaked in oil at every step of the supply chain. Connections like these make it a bit easier to understand that the modern food system uses ten calories of oil energy for every one calorie of food we produce—a topic broached in chapter 6 (Pfeiffer 2006). One can only imagine how many other producers in the global south experienced similar hardships trying to export their produce to Europe.

Thus, to argue that we can characterize our civilizational order as an uneven, yet interconnected and short-lived, petro-market civilization is also a way of claiming that human history can be divided into three major eras: (1) the age of efflorescences, (2) the age of carbon energy and (3) the postcarbon energy age. This periodization and the rationale for it are discussed in greater detail in chapter 2.

The Universalization of Capitalization and the Extension of the Capitalist Mode of Power

The second argument this book makes is that without the discovery and use of fossil fuels—mainly coal and oil—the architecture and magnitude of capitalization, as well as its quantitative and qualitative breadth, depth and geographic scope, would have remained severely circumscribed because it was tethered to renewable energy like wind, water and wood, as well as human and animal power (Di Muzio 2014; Smil 1994). Dominant forms of high-energy–intensive social reproduction—what I will call the carbonization of everyday life—and hierarchical forms of world order are largely the result of exploiting and consuming fossil fuel energy, and this was largely imposed from above *against resistance* to early industrial capitalists and colonial merchants and administrators. Moore captured it best when he wrote that:

> On this score it is well to recollect that there is no evidence that the mass of the population anywhere has wanted an industrial society, and plenty of evidence that they did not. At bottom all forms of industrialization so far have been revolutions from above, the work of a ruthless minority. (Moore 1974, 506).

In this sense, although industrialization was a crucial development, it was a process thrust upon the lower orders of society by 'a ruthless minority' for their own benefit, regardless of the scientific and material benefits that were eventually experienced by more populations that struggled to obtain them. Traditional and mainstream accounts of the Industrial Revolution and the 'rise of the West' more than often obscure this power relationship along with the 'environmental and social costs' of the Industrial Revolution and 'the global inequalities incorporated into the current energy regime' (Barca 2011, 1309). Here we should also be careful not to equate capitalism with industrialization as many have mistakenly done since Marx. When capitalism is understood as a mode of power rather than a distinct mode of production, what becomes apparent is that the relationship between owners and nonowners, the sabotage of human creativity, organized violence and the ritual of capitalization are capitalism's most paramount features. Although Braudel had no *express* theory of capital as the commodification of differential social power, nor a deep understanding of capitalization, in his three-volume study, he did, I would argue, manage to tease out the essence of modern capitalism. As Braudel observed and cautioned: "On a world scale, we should avoid the over-simple image often presented of capitalism passing through various stages of growth, from trade to finance to industry—with the mature industrial phase seen as the only true capitalism. In the so-called merchant or commercial capitalism phase, as in the so-called industrial phase, the essential characteristics of capitalism was its capacity to slip at a moment's notice from one form or sector to another, in times of crisis or of pronounced decline in profit rates" (1983, 433).

Unlike earlier forms of factory capitalism where owners were identified with their products and had little to no investments in other companies and

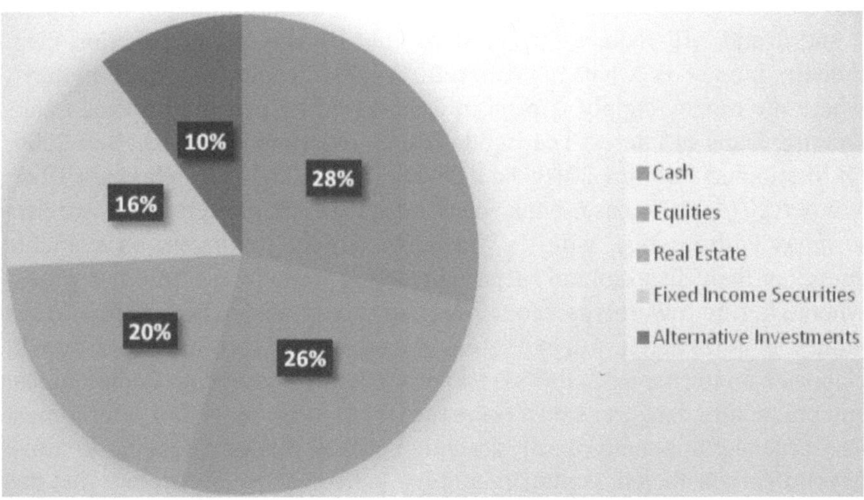

Figure 1.0. Asset class

often oversaw day-to-day activities of the firm, the crucial demarcation of modern capitalism, as Veblen observed, is absentee ownership over a portfolio of income-generating assets. For example, as figure 1.0 demonstrates, dominant capitalists—or what the financial services industry calls high-net-worth individuals (HNWIs)—who have at least US$1 million in investable assets, typically hold their wealth across a range of investments (Capgemini and RBC 2013, 16; Di Muzio 2015a).

The Monetization of Energy and the Capitalization of Debt-Money Creation

A third argument presented in this book is that we need a better understanding of how energy has been monetized and how money is created in modern capitalism (Di Muzio and Robbins 2015). Too often, scholars of IPE and other disciplines compartmentalize knowledge and arguments so that events, moments and analytical categories are looked at separately rather than perceiving them as interconnected phenomena that can help us understand capitalist sociality in new, different and hopefully more enlightening ways. Save some minor exceptions (Alam 2009; Douthwaite 2010 in Heinberg and Lerch 2010; Hall and Klitgaard 2012), most scholars who study money and finance do not study energy and energy problems, and most scholars who study energy and energy problems do not study money and finance. For this reason, I argue here that IPE needs to have a clear conceptualization of how the exploitation of energy—particularly from fossil fuels—and the production and allocation of money function in and across the global economy. This is covered more in depth in chapters 2 and 3 and will require further study, but by way of introduction here, modern money is what Ingham (2004) calls 'capitalist credit money', where the money supply is primarily increased by privately owned banks creating loans as interest-bearing debt and government deficits (Bell 2000; Di Muzio and Robbins 2016; Jackson and Dyson 2013; Rowbotham 1998; Werner 2014).[18] At base, bank loans capitalize the capacity of borrowers to repay with interest, with their capacity largely contingent on available energy at their disposal and success in selling something.[19] In other words, where we find low-energy economies, we should also expect to find lower money supplies and less debt relative to large energy consumers. As we will elaborate on in chapter 2, this system of monetary creation is socially unjust and practically dangerous on a finite planet. As such, monetary reform must be a crucial component of any strategic political platform seeking to move beyond carbon-fueled capitalism and the petro-market civilization that has been constructed and reconstructed since at least 1750.

Hegemonies of Historical Capitalism and Energy Regimes

A fourth argument made in this work is that the literature on shifts in international power since the emergence of carbon capitalism have significantly downplayed, if not totally ignored, the role of energy as a key explanatory factor in the emergence and decline of great powers and the constitution and reconstitution of world order. The major exceptions are Podobnik (2006), Phillips (2006) and Hall and Klitgaard (2012), whose work provide useful discussion but given their reading of capital, have points of departure and arrival different from my own.

Generally speaking, there are two strands within IPE when it comes to thinking about shifting world order. The first is what Arrighi (1994) called 'the hegemonies of historical capitalism', which broadly seeks to understand the historical transition to new hegemonies and the transformation of world order under their leadership. This approach typically has a Gramscian influence that conceptualizes hegemony as a mixture of coercion and consent, with consent to domination more common. These approaches also have a more or less strong cultural component in explaining the acquiescence of the working class to capital and less powerful states to the leading capitalist state with more material capabilities. Arrighi's works (1994, 1997) are perhaps the most prominent and detailed accounts in this research tradition (but see Cox 1987; Gill 1991, 2008; Overbeek 1990; Robinson 2004; Rupert 1995; van der Pilj 1984). Arrighi's main argument is that there is a contradiction between relatively fixed political space and the endless accumulation of capital. Eventually (perhaps 'inevitably' is more appropriate), this contradiction leads to major and prolonged wars that breed 'systemic chaos'. Systemic chaos and social conflict are the fertile grounds from which new hegemons spring up: the United Provinces after the Thirty Years' War; Britain after the American and French Revolutions and the United States after World War I, the Soviet Revolution, the Great Depression and, finally, World War II. Each successive hegemonic power reorganizes the world system anew and seeks to project its particular interests as the universal interests of the world community. Not surprisingly, Arrighi remained true to this general schema and predicted that the War on Terror would largely empower China as the United States became more and more indebted to foreigners to help pay for the war. As we will see, this was a gross misunderstanding because of the failure to realize the importance of energy (Li 2007).

The second major strand issues from a realist perspective and is known in the literature as 'hegemonic stability theory'. In this interpretation of the rise and fall of 'great powers', hegemony is largely understood as dominance, and international power is based on 'material capabilities'. The theory originates with Charles P. Kindleberger but in IR/IPE is most conspicuously associated and developed with the work of Robert Gilpin (1975), who argued that an open liberal economic order and international stability were more likely when one state achieves preponderant power. The hegemon pursues its own national interests, but in doing so also provides desired international public goods, such as a stable monetary order for international trade and military security. Over time, however, this peculiar propensity to self-sacrifice for the greater good of the system, combined with political ineptitude, unsustainable military spending and the rise of new economic powers, leads to relative decline and international instability until, it is presumed, a new liberal hegemon achieves dominance and reorders the system anew.[20]

Although both strands are insightful in their own way, they are also historically inaccurate because they fail to realize that the liberal international economic order and the rise of global capitalism more generally was founded on the *nonrenewable* energy provided by the transition to fossil fuels. As some studies have already remarked, Dutch power was founded on the mastery of wind and the use of peat (an inferior fossil fuel), British power was founded on the exploitation of coal and the power of the United States came from coal and massive discoveries of oil (Dezeeuw 1978; Li 2007, 2008; Mulligan 2010; Phillips 2006; Podobnik 2006). At first glance, this may seem like a minor point to make, but it matters a great deal to how we theorize historical structures and the present conjuncture and think about the likely future trajectory of social reproduction and world order.

Indeed, in both versions we get the sense that the rise and fall of global hegemons is the meta-pattern of human history. Instead of waiting for Godot, we are always waiting for the old power to decline and new powers to arise out of the ashes of systemic conflict and/or military hubris. This is intellectually and practically risky. As Bhambra argues in *Rethinking Modernity*: "[T]he ways in which we understand the past are crucial to our understandings of ourselves and the world in which we live today and . . . if our understandings of the past are inadequate it follows that our grasp of the present will also be inadequate" (2009, 2). One example of this wrongheadedness comes from Arrighi's (2007) argument that international power will shift to China and that its notoriously corrupt Communist Party will give birth to a 'commonwealth of civilizations'. Such claims can only be made by ignoring China's massive energy constraints and the role of the dollar in oil pricing, the size of its population and income levels, its internal contradictions, US military power and the vast ecological degradation caused by its 'modernization'. Like all other nations who have 'modernized' through industrialization, China's leadership is daily making its population more and more reliant on a nonrenewable source of energy while a tiny class of millionaires and billionaires benefits disproportionately through its ownership and development of China's fossil-fuelled development—a pattern begun in Britain with coal (Li 2008; Malm 2012). Reconnecting the rise of global capitalism and hegemonic powers to energy is explored in greater detail in chapters 3 and 4, and the consequences of taking energy seriously are the subjects of the last two chapters.

The Power of the Oil and Gas Industry

A fifth argument in this book is that the power of the oil and gas industry has been the most significant sector of dominant capital shaping and reshaping the globalized social reproduction of a more ubiquitous, yet hierarchical, petro-market civilization. Although the industry is in no way all powerful and has had a number of setbacks and misfortunes, the evidence for its differential power to shape the social process is the differential capitalization of the industry. To some extent, this power is self-perpetuating insofar as the reproduction of a petro-market civilization *requires* both growth and carbon energy due to choices made about the human-built environment and the way in which money creation is capitalized and expanded through interest-bearing loans and state deficits. Currently, the oil and gas industry, as well as their investors who profit from the sale of oil and gas, are locking us into a radically dangerous, unsustainable path that will make any energy transition to renewables and a simpler but arguably more humane pattern of existence far more difficult, if not impossible, for the world as a whole (Di Muzio 2012; Heinberg 2003; Moe 2010). It is also a path that may keep what Brand and Wissen (2013) have appropriately called the 'imperial mode of life' going for a time, but one where runaway climate change will be virtually inevitable. As the former head of NASA's Goddard Institute for Space Studies, James Hansen and colleagues concluded, 'burning all fossil fuels would threaten the biological health and survival of humanity, making policies that rely substantially on adaptation inadequate' (2013, 25).

At the moment, there is no indication that the oil, gas and coal industry will relent in its pursuit to monetize the destruction of the biosphere through the sale and combustion of ever more carbon energy. There is little doubt that almost everyone who lives a high-energy–intensive lifestyle is implicated, but the difference is in the differential power to shape and reshape the terrain of social reproduction, and we should not fool ourselves that noncorporate forces have decisive force in this regard. Investors in oil, coal and gas companies are effectively capitalizing the power of the industry to render the planet uninhabitable for future generations (Di Muzio 2012). This argument is explored in greater depth in chapter 5.

War Made the Fossil Fuel Industry, and the Fossil Fuel Industry Made War

My penultimate argument is that ruling-class warfare played a pivotal, if not decisive, role in intensifying the use of carbon energy. The development of coal and later the oil and natural gas industry, combined with the logic of differential accumulation among the ruling class of owners, essentially created modern warfare and the means to destroy all life on earth. Dominant property owners made the key decisions, never the class of nonproperty owners. What this suggests is that there would be no mechanized total warfare without carbon energy and warring ruling classes competing for the accumulation of money, just as there would be no space exploration or international satellites without carbon energy–inspired science. However, if ruling-class warfare played a pivotal role in intensifying the use of carbon energy and therefore stimulated an entire industrial complex of fuel extraction, steel production and weapons making, we could also make the claim that thereafter the carbon energy industry had an intimate and arguably inseparable interest in war. The fact that the Pentagon is the largest consumer of fossil fuels on the planet is already a key indication of these historically rooted interconnections.[21] I will explore this argument further from chapter 3 on.

The General Crisis of Social Reproduction

My final argument is that with the peak of global fossil fuels, we will not simply be witness to 'peak globalization', 'peak trade' or the end of economic growth, but what I have called a general crisis of social reproduction (Curtis 2009; Di Muzio 2011; Heinberg 2011; Rubin 2012). What I mean by a general crisis of social reproduction is a multiscalar, multidimensional and internationally interconnected situation whereby current patterns of energy-intensive production, consumption and reproduction can no longer be sustained. This will herald what I call a second great transformation in civilizational order. If the first great transformation was the metamorphosis of agrarian societies of low growth into more urban, market-dependent societies with compound growth and some form of democratic planning, then the depletion and greater cost of fossil fuels over the coming decades will signal the unevenly experienced creeping end of petro-market civilization in the twenty-first century. There are a number of incredibly important dimensions to this crisis that will be unpredictable, but I will explore some likely outcomes and tendencies in more detail in chapter 6. I now turn to the organization of this book.

ORGANIZATION OF THE BOOK

In chapter 2, I investigate how the tradition of political economy has understood energy and note that after Marx and Jevons, and with oil becoming more common as a surplus energy source in the early twentieth century, energy was all but ignored in the political economy literature until the oil price shocks of the 1970s. I then explain why energy source, supply and availability should be central to theorizations in political economy by providing a brief excursus on energy. I then investigate how political economy has theorized historical time and argue for a new periodization premised on the energy base of social power relations: the age of efflorescences, the age of carbon energy, and the post-carbon age. I then provide an overview of my theoretical approach grounded in the framework of capital as power and explain how this differs from prevailing explanations of capitalism as a specific mode of production. The final section explains the linkage between the three major concepts that will be used throughout this work: capitalization, energy and social reproduction.

Chapter 3 opens by providing an overview of prevailing accounts on the transition to capitalism. In the later sections, the chapter moves to demonstrate that they are not convincing because virtually all of them fail to take the transition to fossil fuel energy situation seriously. I begin to offer a new theorization by considering the limits to surplus accumulation writ large before the transition to carbon capitalism. The age of efflorescences might have experienced what we now call economic growth, but it was always short lived and never sustained. This is because the energy sources used were limited to wood, water, wind, peat and the animate power of humans and certain domesticated animals. These fuel sources tied societies to the rhythms of photosynthesis and the natural environment limited their capacity, regardless of culture. However, a confluence of events transpired in England that led to the carbon age proper. I argue here that coal energy is not a sufficient explanation for the transition to capitalism in England, but it is a vital part of the story and the origins of our more globalized petro-market civilization today.

In England, unlike in other stock markets of the world at the time, we witnessed an unprecedented rise in companies and their levels of capitalization, not to mention the debts of nation-states listed on the London Stock Exchange. This would not be surpassed until the United States began to exploit a new form of carbon energy—oil—in massive and sustained quantities at the turn of the twentieth century. In the next section, I argue that the British Empire could be conceived of as the Empire of Coal by considering coal usage, industrialization, colonialism and wrenching dislocations of the social fabric. The final section briefly considers Britain's military-led transition to oil.

Chapter 4 considers the rise of the petro-market civilization in the United States where coal was not only used as an energy source early on, but also where oil was found in astronomical quantities. Indeed, the United States was the first Saudi Arabia of oil, but also had considerable other energy stores such as wood and coal. I open the chapter with a discussion of the revolutionary struggle to found a new nation separate from the British Empire and how the Constitution ensured private property, slavery and an emergent capitalist class. I then move to discuss slavery and capitalism in the United States and how the Civil War (1861–1865) contributed to increasing capitalization in the United States, mainly through federal debt.

This section also considers how ownership was increasingly centralized and concentrated, along with the political discourse on the need for expansion and foreign markets at the turn of the century, when the frontier of the continent was believed to be closed. I then examine the relationship between energy, capitalization and social reproduction in the United States as oil transformed from a source of light (kerosene—the product Standard Oil's John D. Rockefeller used to become the world's first billionaire) to a primary source of motor fuel (gasoline and diesel). This paves the way for an investigation of the United States as an Empire of Oil and the global consequences for its social reproduction.

Chapter 5 considers three important dimensions of global carbon capitalism. In the first section, I provide an analysis of what the *Financial Times* labelled 'the new seven sisters', altering Enrico Mattei's earlier formulation that focused solely on the international oil companies. State-run oil companies are now far more important to the global economy and to some extent, they have shifted global power relations and are likely to do so in the future as oil prices increase (see also Victor et al. 2012). These new relations are also briefly considered within the context of the growing demand for energy among the BRIC countries (Brazil, Russia, India and China).

The second dimension of this chapter highlights and reconsiders the role of petro-dollars, redistribution and the idea of a perpetual war. I empirically demonstrate how a bank–oil–armament nexus could be said to be profiting off the prospects of perpetual war and show how this is related to their differential accumulation. Perhaps the most controversial aspect of this section is the suggestive argument that the Volcker shocks were used as a strategic weapon to ensnare developing countries and working classes in unending debt. The final section considers the prospects for a renewable post–carbon energy future by analyzing the capitalization of the oil and gas sector and the renewable energy sector. As the reader will see from the empirical details, the world is nowhere near ending its high-energy forms of social reproduction and world order based on fossil fuels.

Chapter 6 closes this study by considering six dimensions that seem critical to the expanded reproduction of carbon capitalism and a worldwide petro-market civilization: food, science, health and population, employment and mobility, climate change, global capitalization and democracy and civil peace. By assessing these dimensions, I argue that we are daily being locked into an unsustainable pattern of development that will likely lead to a general crisis of social reproduction. In the coda to this chapter, I discuss what I call the Douglasian thesis that 'power concedes nothing without a demand' and combine this with what I call the Carr–Diamond thesis that understands how elites can actually benefit from harming others and the biosphere.

NOTES

1. *Oil Shockwave: Oil Crisis Executive Simulation*. Final Report. 2005. http://www.secureenergy.org/policy/oil-shockwave-2005-report. A second scenario was run in 2007: *Oil Shockwave: Oil Crisis Executive Simulation*. Final Report. 2007. http://www.srwolf.com/reports/OS_2007_Report_042808.pdf. See also John M. Broader (2007), 'A War Game Supposes Scarce and Risky Oil', *New York Times*, November 2. At the time of this writing, current stocks of oil in the SPR are 695.9 million barrels. At current consumption rates, this is roughly enough for about forty days' worth of oil in the United States. http://energy.gov/fe/services/petroleum-reserves/strategic-petroleum-reserve#Current.

2. Using BP's 'Statistical Review of World Energy of June 2013', the average price of a barrel of oil was US$22.54 from 1984 to 2004. However, by 'cheap', we mean only in relative prices and major currencies. Given the subsidies to the oil industry and the environmental damage inflicted on the biosphere and communities, oil is far from 'cheap'. I thank Matt Dow for reemphasizing this point to me.

3. Jevons's concern over British coal depletion could be said to start the tradition for fossil fuels, but he is undoubtedly influenced by Malthus, who worried about food energy.

4. Crosby refers to it as 'buried sunshine'.

5. Data are from *BP Statistical Review* 2013.

6. Canada, Denmark and Norway are exempted. As of 2011, members hold about 4.1 billion barrels of oil in their stockpiles (IEA 2012, 7).

7. The reboot of the *Mad Max* franchise after lying dormant for over thirty years may be a sign of the times.

8. Hobson (2013a, 1030–31) puts the emergence of the field back in the mid-1700s with the birth of classical political economy. His main contention is that there was never any separation of 'politics' from the 'economy' in 'political economy'. He argues that IR scholarship from the 1940s to the late 1960s ignored that economic dimensions should be understood as a giant historical aberration. Although Hobson certainly has a point that IPE did not have a virgin birth in the 1970s and even today is informed, implicitly or otherwise, by the classics, the institutionalization and

professionalizing of the field can safely be placed in the 1970s, accelerating in the following decades due largely to the expansion of universities, journals and books, not to mention affordable oil to sustain growing economies and ever more academics.

9. It should be noted that the authors were primarily faulting Marxist thought.

10. Calculations are made from BP's *Statistical Review of World Energy 2014*: http://www.bp.com/en/global/corporate/about-bp/energy-economics/statistical-review-of-world-energy.html.

11. To minimize the amount of lengthy citations, these sources are cited in chapter 2 in the section on political economy and energy.

12. Unfortunately, the study by Gill and Law has not been updated since its original publication.

13. The seminal formulation is 'fossil capitalism' and originates with Altvater (2007), but I prefer carbon capitalism, given that oil and coal are 75 percent to 87 percent carbon (depending on the type) and that most natural gas is methane made of 75 percent carbon. The chief difference between Altvater's conceptualization and my own is my non-Marxist focus on capital as a mode of power and capitalization as the dominant ritual of capitalist societies, discussed in greater detail in chapter 2. As I was researching and writing this book, Urry's book (2013) appeared and introduced the term carbon capital. However, 'capital' is never analytically defined, and he does not approach fossil fuel–based societies from the capital-as-power perspective as I do in this study.

14. Governments are capitalized through the private ownership of their national debt. We discuss this in greater detail throughout this study, as it is integral to the rise of capital as finance and only finance.

15. I do not pretend to be able to detail and explain everything and, like all books, there will be gaps and silences. As academics trying to understand limited parts of our universe, this is inevitable. I can only say that I will be grateful to those who point out the shortcomings in this study in constructive debates or comments.

16. *Reuters*, 'Green Groups Point to Ash Cloud Silver Lining', 21 April 2010.

17. Jeffrey Gettleman, 'With Flights Grounded, Kenya's Produce Wilts', *New York Times*, 19 April 2010.

18. See also Martin Wolf, 'Strip Private Banks of Their Power to Create Money', *Financial Times*, 24 April 2014, and David Graeber, 'The Truth Is Out: Money Is Just an IOU and the Banks Are Rolling in It', *The Guardian*, 18 March 2014. *97% Owned* is also a telling documentary.

19. Polanyi (1957, 41) reminds us that in a market economy: "All incomes must derive from the sale of something or other, and whatever the actual source of a person's income, it must be regarded as resulting from sale."

20. Hobson (2013a, 2013b) calls out the contradiction in thought brilliantly.

21. Andrew Herndon, 'Pentagon World's Biggest User of Fossil Fuels, Shops for Biofuels', *Bloomberg News*, 19 October 2012, http://www.businessweek.com/news/2012-10-19/gevo-syntroleum-vying-for-military-biofuel-supply-deals.

2

The Political Economy of Petro-Market Civilization

As Smil has argued, '[E]nergy is the only universal currency' (1994, 1). If the world's progeny are able to look back centuries from now, the petro-market civilization constructed during the age of carbon energy will look like a small, but not insignificant, blip in human history—a currency spent within a few centuries and in profoundly harmful and wasteful ways. To the best of our knowledge at the present time, anatomically modern humans first appeared about 150,000 to 200,000 years ago in Africa. Likely due to rapid population growth, behavioral changes and competition for nourishment, the great migration out of Africa began around 60,000 to 130,000 years ago (Bae et al. 2014; Mellars 2006). From there, and for reasons that are still debated, *Homo sapiens* replaced other members of the genus *Homo* that had settled in Eurasia (McKie 2013). After the mastery of fire, the main transformation recognized in human sociality was the blending of hunting and gathering with the domestication of plants and animals known as the Neolithic or agricultural revolution (Ehrlich and Ehrlich 2008; Weisdorf 2005). Dated 10,000 to 12,000 years ago, this revolution in human social reproduction is believed to have first occurred in the Fertile Crescent and later spread by way of colonization as native populations practicing different forms of social reproduction were devastated by settler violence, alcohol and, in many cases, new diseases (Diamond 1997). Settled agriculture gave rise to the first cities and more apparent class hierarchies than witnessed in groups of hunters and gatherers, where a rough egalitarianism is thought to have prevailed (Boehm 2001; Diamond 1997; Fernandez-Armesto 2000; Price 1995). It is still widely debated why seemingly well-nourished hunters and gatherers transitioned to farming, given that the practice of domesticating plants and animals was 'back breaking, time consuming, and labor-intensive' (Weisdorf 2005, 562). But Manning suggests perhaps the most plausible answer for the historical riddle:

> Farming did not improve most lives. The evidence that best points to the answer, I think, lies in the difference between early agricultural villages and their pre-agricultural counterparts—the presence not just of grain but of granaries and, more tellingly, of just a few houses significantly larger and more ornate than all the others attached to those granaries. Agriculture was not so much about food as it was about the accumulation of wealth. It benefited some humans, and those people have been in charge ever since (2004, 38).

We may never know for certain whether the transition to agriculture was the result of a power process, and farming was almost certainly synchronic with hunting and gathering. But the fact that slavery and other forms of human bondage emerged at the same time as agriculture and cities is highly supportive of Manning's thesis (Drescher and Engerman 1998; Heuman and Burnard; Nikiforuk 2012). Indeed, even by 1772, the British agricultural writer Arthur Young (1741–1820) estimated that of a world population of 775 million, only 33 million could be considered in any way 'free'. The remainder, some 742 million, existed in various forms of servitude to the 4.3 percent of the global population living as dominators in greater states of freedom (Nikiforuk 2012, 12). Although certainly more complicated than the exploitation of a new energy source, Nikiforuk argues that the *coup de grâce* for world historical slavery was the transition to coal and later oil as primary energy inputs for production in industry and agriculture. What this suggests is that without the energy from fossil fuels, the naturalization and normalization of various forms of human servitude and slavery would not have been overcome for a large swathe of humanity—regardless of religious or moral sentiments. Indeed, most slave holders understood that their lifestyles, privileges and freedoms were wholly contingent upon slave labour, and until there was an alternative or they were compensated for their 'property', they fought tooth and nail to safeguard the institution that garnered their private wealth and personal liberty (Blackburn 2011; Davis 2006; Draper 2010). In deeply contradictory ways, what we could call the fossil fuel revolution altered the course of human history and ushered in new conditions of existence and domination, as well as new forms of social reproduction and a world order more reliant on the market and fossil fuels. This sociospatial order is what I have called a petro-market civilization, and to explore its uneven and hierarchical emergence and development is the task of this work. As a starting point for further argumentation, this chapter first discusses the concept of energy before moving on to consider how political economy as a body of knowledge has conceptualized and dealt with the question of energy. The chief argument here is that whereas early political economists such as Marx and Jevons understood the importance of energy to industrial society, once oil became more and more common throughout the heartland of global capitalism, the question of energy was all but ignored by the field until the oil price shocks of the 1970s. After some noticeable debate in international relations/international political economy (IR/IPE) throughout the 1970s and some minor exceptions in the decades that followed, energy was pushed to the margins of the discipline until the so-called War on Terror and the oil price spike of the 2000s. Since energy is central to this study, I begin with a brief excursus on it. In the ensuing sections I consider how political economists have theorized historical time and argue for a new periodization premised on the energy base of social power relations and world order. I then introduce the capital-as-power framework in greater detail and finish the chapter with a discussion of what I call the energy–capitalization–social reproduction nexus.

A BRIEF EXCURSUS ON ENERGY

Although vital for life on Earth and the complexity of life-forms and ecosystems, the fundamental importance of energy is generally taken for granted in everyday social life. According to Smil (2006, 1), the term 'energy' originated as a Greek compound word in Aristotle's *Metaphysics*. Combining 'in' and 'work' to form 'energeia', the concept usually means the capacity or ability to do work. But for energy to be useful, it must be converted in some way. For example, the entire process that underpins life on earth, photosynthesis, sees plant life convert radiant or electromagnetic energy from the sun into chemical energy useful for other life-forms.

Converting energy almost always entails limitations and costs and is therefore never straightforwardly beneficial to humans or other species. We only need to think about the problem of disposing of radioactive nuclear waste and the tons of carbon dioxide emitted from the combustion of fossil fuels to illustrate the point. Although we know that energy comes in many forms (e.g., chemical, thermal, kinetic, electrical), we can make a distinction between renewable sources of energy such as geothermal, wind, tides, waves and sunlight and nonrenewable sources of energy like fossil fuels and uranium. The former can be reproduced on a human time scale, but are flow or rate limited, whereas the latter cannot be renewed at an adequate rate to be of continued use to humanity, as they can be considered stock limited (Gleick and Palaniappan 2010, 11156). For instance, crude oil and coals are believed to be produced from high temperatures and pressure interacting with organic matter over hundreds of millions of years (Braun and Glidden 2014; Smil 2006, 105). As stated earlier, anatomically modern humans have only been around for about 150,000 to 200,000 years. Although there is no telling exactly how long our species will last, there is little sense imagining a second oil age hundreds of millions of years from now.

As a series of books and studies have demonstrated, outside of global climate change, the major long-term problem faced by humanity is the nonrenewable nature of oil, coals and natural gas, given that the social reproduction of a high-energy–intensive civilization is dependent upon them (Curtis 2009; Deffeyes 2005; Di Muzio 2011; Heinberg 2003, 2011; Hirsch et al. 2005; Rubin 2012). Although we have known about the potential for shortages and price fluctuations for many decades, precious little has been done to transition society away from carbon energy–backed development projects and toward renewable alternatives (Di Muzio 2012). As we will discuss in greater detail, the reasons for this have mostly to do with the power and profit of the few and the cultural hegemony produced by high-energy forms of consumption. But what we can note here is that our consumption of carbon energy has barely decreased from 1973 to 2011 as figure 2.0 makes clear. Moreover, the world's total final consumption of energy has increased from 4674 million tons of oil equivalent (MTOE) to 8918 MTOE, with fossil fuels making up 66.4 percent of the world's total final consumption in 2011 (IEA 2013, 28). Total primary energy supply is also dominated by fossil fuels, at 81.6 percent, as of 2011 (IEA 2013, 6).

To be sure, there are still low-energy communities that experience little to any benefit from the fossil fuel revolution, and those communities will have comparatively lower rates of growth, lower money supplies and lower valued currencies. Their overall debt—state, business and personal—will also be lower compared with high-energy–intensive political economies because money is created to capitalize capacity and with low energy consumption, there is typically lower capacity to perform work and therefore less to capitalize for profit. As the United Nations Development Programme (UNDP) reminds us, although energy availability is not a sufficient cause for 'devel-

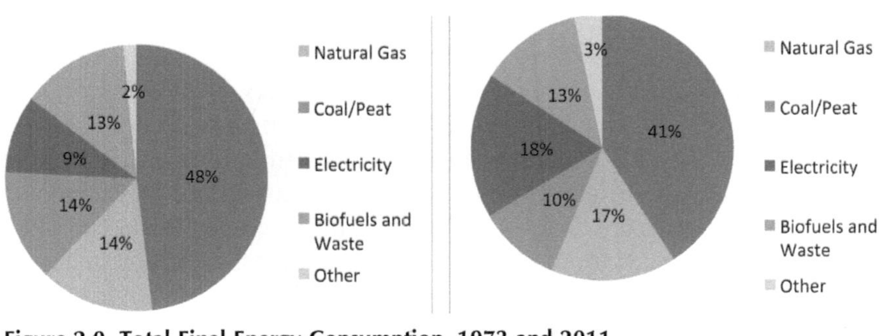

Figure 2.0. Total Final Energy Consumption, 1973 and 2011

opment', it is certainly a necessary and vital one (UNDP 2000, 41). Indeed, the UNDP notes how:

> Energy services are a crucial input to the primary development challenge of providing adequate food, shelter, clothing, water, sanitation, medical care, schooling, and access to information. Thus energy is one dimension or determinant of poverty and development, but it is vital. Energy supports the provision of basic needs such as cooked food, a comfortable living temperature, lighting, the use of appliances, piped water or sewerage, essential health care (refrigerated vaccines, emergency and intensive care), educational aids, communication (radio, television, electronic mail, the World Wide Web), and transport. Energy also fuels productive activities, including agriculture, commerce, manufacture, industry, and mining. Conversely, lack of access to energy contributes to poverty and deprivation and can contribute to economic decline. (UNDP 2000, 44)

In other words, a key component—but certainly not the only component—of global poverty is energy deprivation. Poor people cannot command anywhere near the energy resources consumed by the global affluent. The UNDP defines energy poverty as 'the absence of sufficient choice in accessing adequate, affordable, reliable, high-quality, safe and environmentally benign energy services to support economic and human development' (UNDP 2000, 44). The UNDP estimated that 1.7 billion people on the planet are without the benefits of electricity, and another 2 billion cook with unsafe cooking fuels. What this suggests is that a key dimension of world order and international relations is the radically *uneven* access to and consumption of energy. It also brings our attention to the fact that for energy to be useful, it must be affordable, accessible and abundant. Thus the source, quantity and quality of energy supplies, as well as methods used to store, distribute and convert energy, are of critical importance for the evolution and social reproduction of human societies. As Smil summarizes, '[F]rom the perspective of natural science, both prehistoric human evolution and the course of history may be seen fundamentally as the quest for controlling greater energy stores and flows' (Smil 1994, 1). This brings us to another key concept we should be familiar with if we want to understand the political economy of energy: ERoEI.

ERoEI stands for energy returned on energy invested and is calculated by taking the amount of acquired useful energy and dividing it by energy expended to acquire it, or expressed as an equation: ERoEI = Acquired Useful Energy ÷ Energy Expended. Although calculations are likely never exact, given the range of energetic processes that might be involved in extracting more useful energy, at a minimum, the equation does give us some indication of net energy gain, a breakeven point or net energy loss. Thus, over time it would make little sense to use ten units of energy to acquire five units of useable energy. A related issue is the economic cost of obtaining net energy gains. For example, the tar sands of western Canada have been known about for some time, but were not economically profitable to exploit until oil reached about US$80 per barrel (Chilingarian and Yen 1978: Nikiforuk 2010).[1] It was only in 2003 when Canada's proven oil reserves ballooned from 10 billion barrels to 180 billion barrels 'after oil sands resources were deemed to be technically and economically recoverable', making Canada the country with the third-largest proven oil reserves in the world (EIA 2014). Despite the environmental and ecological consequences of transforming tar sands into useful energy, such as runaway climate change, poisoned water supplies, destruction of communities and deforestation, the Energy Information Agency (EIA) of the United States reports that Canada will increase production from 4 million barrels per day (bbl/d) to 6.7 million bbl/d by 2040.

In addition to ERoEI, another concept called Jevons Paradox is worth mentioning. William Stanley Jevons, who we will encounter in the next section of this chapter, was an English economist famous for divorcing politics from economics by mathematizing 'the economy' and for inspiring the marginal revolution in economic thought. In *The Coal Question* (1865), Jevons observed that increases in technological efficiency increased rather than decreased the rate of energy consumption. In other words, the more technologically efficient societies become at extracting and using resources (including energy), the greater the consumption rate of that resource (Alcott 2005; Alcott in Polimeni et al. 2008). For example, people may tend to drive more than they might otherwise do with greater fuel efficiency, thereby increasing the rate of gasoline consumption over time as everyone drives more. However, Jevons Paradox is not an iron-clad law that can be applied with equal weight across all political communities, given the unevenness of energy consumption. Moreover, in its extreme form, the paradox suggests that striving for energy efficiency or greater energy conservation is doomed to fail since we will only end up encouraging more ruinous consumption. There is some considerable truth to this position, but increasing efficiency and avoiding unnecessary waste can also be considered worthwhile goals if they are combined with other policies that discourage needless consumption and increase the durability of goods (Wackernagel and Rees 1997).

Thus, to sum up, it is worth keeping the following key points in mind as we continue our investigation into the emergence and development of carbon capitalism:

- We are, at base, a solar economy reliant on plants that convert the sun's radiant energy into useful chemical energy by photosynthesis.
- Energy is broadly conceived as the capacity to do work.

- Energy must be converted for it to be useful.
- Broadly conceived, there are two sources of energy related to a human time scale: renewable and nonrenewable.
- Total final energy consumption has been increasing and is largely the product of our use of and combustion of fossil fuels, which continue to make up 66.4 percent of the world's total final energy consumption.
- Energy consumption is radically unequal across societies, and energy poverty is a key, though certainly not the only, reason for global poverty.
- ERoEI, although often difficult to calculate with complete accuracy, can at least provide us with an approximation of net energy gains, a break-even point or the potential for net energy losses.
- The rate of consumption of energy (or a resource) *may* actually increase as society becomes more technologically efficient.

Now that we have introduced how energy has been conceived of in the natural sciences and introduced some of the key concepts related to energy and society, we are in an ideal position to consider how political economy has dealt with the question of energy.

POLITICAL ECONOMY AND ENERGY

As a body of knowledge, political economy has had many points of contact with the concept of energy but has, perhaps strangely, never offered a deep theorization of its importance to civilizational order and social change until the work of Debeir, Deléage, and Hémery (1991).[2] I say 'strangely' because one of the fundamental problematics of early political economists was to uncover the source or origins of 'surplus', or what economists today call economic growth. As we will discuss in a moment, the slight exception here is William Stanley Jevons (1865), who understood that British commercial supremacy, along with its material standard of living and martial power, was founded on a bedrock of nonrenewable coal.

The body of knowledge known as classical political economy emerged in the seventeenth century and took as its problematic the nature, source and distribution of wealth (Aspromourgos 2005, 2–3; Milonakis and Fine 2009). Marx argued that William Petty was the founder of classical political economy, and according to Aspromourgos, Petty originated 'the concept of an economic or social surplus' (2005, 1 and 12). For Petty, the source of a surplus resulted from the difference between worker output on the land and their consumption. As long as output exceeded consumption, a surplus could be generated to create exchange value or money and private profit—Petty's chief concern. Petty was writing at a time when the discourse of agricultural improvement pervaded much of the southern English landlord class, and Petty himself was enriched by gaining title to confiscated lands in Ireland—lands expropriated by Cromwell's capitalized conquest of Ireland (Di Muzio 2015a, 116ff; Wood 1984). Although Ireland can be considered the first foreign laboratory for English agricultural improvement, the discourse of improvement for private profit served as a key justification for additional colonial expropriations as the British Empire extended its global reach in the following centuries (Wood 2002, 164).

The idea that land and labour were the primary ingredients for generating surplus or growth continued on in political economy until 'stock' and much later 'capital' were added as factors of production deserving of their own rewards (at least in neoclassical theory). However, as Wrigley has pointed out, the major political economists of the classical era in England—Smith, Malthus and Ricardo—did not understand growth (or wealth) as something that could be prolonged and constantly amplified. Consequently, there was no hope that the labouring poor would eventually see their standards of living rise progressively. They were condemned by their social status and the limits of the natural world to a perpetual hierarchy of class relations where the wealthy and powerful appropriated the bulk of the surplus. The primary reason underpinning this belief was that the amount of land available for productive use was limited (Wrigley 2010, 10).[3] Wrigley also makes the claim that since the majority of the population consisted of poor labourers, no market for mass-produced goods beyond the 'necessaries of life' would have been encouraged—aggregate demand would have been too low and too narrow (2010, 13). So in spite of the fact that all three political economists were well aware of coal use, there was still no fundamental theorization that fossil fuel energy could be one of—if not *the*—crucial factors in sustaining compound economic growth and the accumulation of wealth.

Debeir, Deléage and Hémery (1991, xiii) make the argument that there is something different about Karl Marx's approach to political economy (see also Foster 1999). They argue that unlike the classical political economists, Marx started from the very premise that humans and nature are inseparable in both theory and practice. There is little doubt that Marx understood the transformative practices of capitalist accumulation on the environment and their links to historical forms of social reproduction.[4] However, when it came to his *scientific* analysis of capitalism, Marx pinned his hopes of explaining the accumulation of surplus in the form of money values and the necessity for communist revolution on the fact that labour power was the *sole* source of added value in the production of commodities. He argued that during the labour process, workers add more value to the commodities produced than they were actually paid by capitalists. The problem for Marxism has always been to demonstrate this scientifically: to somehow transform the Marxist unit of labour values into prices and show the math of exploitation at work. Despite some considerable yet flawed attempts, this has not and likely never will be accomplished (Nitzan and Bichler 2009, chapters 6 and 7 for the full critique). The legacy for Marxism was theoretically and practically tragic for a convincing theory of capital accumulation and energy. Debeir, Deléage and Hémery surmised:

> Most Marxists thereafter conceived energy problems only as problems of production and exchange; they made them part of the notion of productive force which, in a context of abundant resources, they used mainly for rhetorical purposes. Energy became one of the main blind spots of Marxist thought. This prolonged drift eventually led to the *irrational conviction* that natural constraints would soon be overcome, the chief credo of nineteenth-century and twentieth-century scientific ideology (1991, xiii; emphasis added).

So although Marxism did not fall prey to the abstractions of neoclassical economics and was firmly rooted in a historical understanding of social property relations and class conflict, its narrow focus on the production of commodities during a period of industrialization blinded it from offering a deeper and more historically informed theorization of energy, social reproduction and the constitution and reconstitution of capitalist world order (see also Hornborg in Strauss et al. 2013, 50–51).[5]

Ironically, it was William Stanley Jevons (1865), one of the men who would do the most to abstract the economy from society and nature in his later work, who noticed the relationship between cheap coal energy, industrial development and British 'progress'. The title of his work in its entirety paints the picture clearly: *The Coal Question: An Inquiry Concerning the Progress of the Nation, and the Probable Exhaustion of Our Coal-Mines*. Calling his era the Age of Coal, Jevons wrote:

> Day by day it becomes more evident that the Coal we happily possess in excellent quality and abundance is the mainspring of modern material civilization. It is the material energy of the country—the universal aid—the factor in everything we do. With coal almost any feat is possible or easy; without it we are thrown back into the laborious poverty of early times. This question concerning the duration of our present cheap supplies of coal cannot but excite deep interest and anxiety wherever or whenever it is mentioned: for a little reflection will show that *coal is almost the sole necessary basis of our material power*, and is that, consequently, which gives efficiency to our *moral and intellectual capabilities*. (Jevons 1866, 5; emphasis added)

Thus, unlike Marx whose theory of capitalist accumulation is not chained to nonrenewable coal power but to the power of workers, Jevons openly questions the relationship between civilizational 'progress' and the energy released from combusting coals. However insightful Jevons' work on the coal question is, he could not see how coal, oil and natural gas would continue to shape and transform patterns of social reproduction and world order dynamics and offered no deep or sustained *historical* theorization of the emergence of capitalism and coal power. In fact, despite being inspired by debates in physics on the conservation of force (later, energy), he largely evacuated a social theory of energy, politics and power from his later thinking in an effort to make economics a science founded on math and utility (Mirowski 1989; White 2004). His true desire was to provide 'a marginalist explanation for the laws of supply and demand' in abstraction from social history and political power (White 1991, 224). This is why I consider Jevons only a slight exception to the tradition of political economy—a tradition that he eventually eschewed in pursuit of economics as a 'science' that could parallel some of the mathematical certainties of physics—the nascent master science of his age.

Since the time of Jevons and the establishment of 'modern economics', the understanding of the economy has become more and more detached from material reality (Robbins 1935). With some recent exceptions, the Marxist tradition largely subsumed energy under the concept of 'mode of production' so that it came to be implicitly treated as 'neutral, unlimited [and] inexhaustible' (Debeir, Deléage, and Hémery 1991, xii; recent exceptions are Abramsky 2008–2009; Altvater 2007; Huber 2008, 2013; Keefer 2010 in Abramsky 2008–2009, 81–9; Magdoff and Foster 2011). The major institutionalists like Veblen, Galbraith, Polanyi and Schumpeter (the latter two are also recognized as being part of the German historical school) also failed to provide us with a framework of political economy that took the relationship between energy and society as fundamentally important for understanding and explaining social change and where we might be headed in the future. Given that more and more energy was coming on line in Western Europe, the Soviet Union and the United States, they can perhaps be forgiven for taking the energy basis of society for granted.

However, the oil price shocks of the 1970s and their aftermath made it blatantly apparent that energy might be important after all. Out of this recognition and other developing factors (see Brown 1973; Hancock and Vivoda 2014; Jones 1981a, 1981b) emerged the field of IPE, largely a result of the dissatisfaction of scholars in the field who thought that leading or mainstream thinking was not well equipped to deal with a rapidly changing and increasingly interconnected world. Inside and outside the IPE literature, some considerable focus was directed toward the importance of oil to the global economy, the impacts of industrialism on the environment and the limits to carbon energy–fuelled growth (e.g., Levy 1979; Meadows et al. 1972).[6]

But this was to be a relatively brief interlude. By the early 1980s, the price of oil plummeted and by the end of the decade, the Soviet Union was in free fall and eventually dissolved. With the Cold War effectively over, IR/IPE largely shifted its focus to a litany of general and specialized debates as the university system continued its expansion and more and more scholarly journals appeared on the scene. The literature of the late 1980s and 1990s is too vast to cover here, but in summary, major debates focused on (1) institutions, regimes and the possibility of overcoming anarchy; (2) globalization as a new phase of capitalist development, (3) the debt crisis in the global south and the Washington consensus; (4) the reemergence or intensification of global finance, trade and the neoliberal restructuring of state–civil society complexes, (5) the resistance to capitalism and the prospects of a new double movement and (6) empire and the new imperialism. As important as these debates are, what they share in common is a general ignorance of the deep interconnections between energy, social reproduction and differential capitalization.

Until the tragic events of 9/11 and the launch of the so-called War on Terror, the small inroads that were made on the importance of (mainly) fossil fuels to the global economy were all but forgotten and pushed to the margins of the field. Starting in 2001 when the ongoing War on Terror was launched, the world was once again subject to mounting oil prices, and by 2007, a global financial crisis that wiped half the value off stock markets around the world (Di Muzio 2014). Moreover, concerns about global climate change, increasing environmental degradation and the limits to growth were renewed (Bardi 2001; Hall and Day 2009; Heinberg 2007; Newell and Paterson 2010; Rubin 2012). In this environment, a litany of new studies—often outside of the IPE field specifically—also started to concentrate on fossil fuels, the political economy of energy and the 'limits of the possible' imposed upon humanity by the fact that they are nonrenewable sources of energy (e.g., Abramsky 2010; Alam 2009; Bina 2006; Clark 2005, Deffeyes 2005; de Graaff 2012; Di Muzio 2007, 2011, 2012, 2014; El-Gamal and Jaffe 2010; Friedrichs 2010, 2013; Heinberg 2003; Hirsch et al. 2005; Huber 2008, 2013; Li 2007; McKillop and Newman 2005; Ovadia 2013b; Parra 2004; Pfeiffer 2006; Rubin 2009; Rutledge 2005; Yergin 2006, 2012).[7]

Given the high historical cost for oil, knowledge that the production of conventional oil has likely peaked or will do so very soon and the fact that coal and natural gas are also nonrenewable, it will be increasingly difficult for political economy and its international cousin, IPE, to ignore the energy base of the global political economy and how this shapes patterns of social reproduction and world order. In this light, drawing on the work of others I propose a new rendering of historical time in the following section.

POLITICAL ECONOMY AND HISTORICAL TIME

As most historians recognize, dividing the human past and characterizing it in one way or another always involves a choice by the scholar. Although there are certainly events that occur at a particular moment in time, thinking diachronically about human development and evolution can be a more difficult task when it comes to periodization. However, if fossil fuels are nonrenewable on a scale useful to humans, then it would appear that carbon capitalism and petro-market civilization are transitory historical structures within what Braudel (1983) called the *longue dureé*. We can thus conceive of three historical eras: an age before fossil fuels, the age of carbon energy and the post–carbon energy age for the duration of human existence. I offer a brief sketch of the first two periods here, and discussion of the post-carbon era is discussed in the final chapter of this work.

I call the age that preceded that of carbon energy the 'age of efflorescences' after the work of Jack A. Goldstone (2002). Building on the work of a growing consensus among historians, Goldstone proposed that the binary thinking between a static and inertial past and the sustained growth of 'modernity' is historically inaccurate. Instead, Goldstone argues that by the early modern period there is considerable evidence to suggest that many political economies were far from stagnant and on the verge of constant crisis. He proposes that what we observe are periods of 'efflorescences' or 'relatively sharp, often unexpected upturn[s] in significant demographic and economic indices, usually accompanied by political expansion and institution building and cultural synthesis and consolidation' (2002, 333). However, what was highly specific about this era of 'organic economies' was not simply Goldstonian 'efflorescences' but the fact that 'economic growth' was never sustained (Wrigley 2010). The most convincing reason why this was so appears to be the fact that exosomatic energy consumption was very low and tied to renewable sources like wind, biomass (mainly wood), water, animal and human capacities. The exploitation and combustion of coal, and later oil and natural gas, severed this historical chain and expanded the limits of the possible for power and social reproduction (Bakker and Gill 2003).

It is somewhat difficult to date the arrival of the carbon energy age and sustained energy growth from a global perspective. For example, the transition to mass coal production and consumption first happened in Britain and the turning point seems to be around 1750 when production virtually doubled from the first decade of the eighteenth century and total energy consumption increased from 169 to 231 petajoules (Wrigley 2010, 37). The rest of Europe also accelerated their production and consumption of coal by 1830 (Podobnik 2006, 28). For the rest of the world, the transition began much later, with China and India accelerating their consumption yearly since 2000. Although coal is unevenly consumed across the world, according to the IEA, coal continues to have a bright future: "[C]oal use has never stopped increasing and the forecasts indicate that, unless a dramatic policy action occurs, this trend will continue in the future".[8] If we consider a little 'oil and natural gas age' within the greater age of fossil fuels, then it is comparatively small and begins in the United States starting from the mid-1800s, but only outpaces coal as the world's most consumed fuel sometime in the early 1960s (Smil 2006, 119). The difference can also be seen in the differential capitalization of these energy sources. The global coal industry has a total market value of US$ 115 billion, whereas the capitalization of the largest oil and gas companies (not including the estimated market value of state-run oil firms) was US$ 3.1 trillion in June 2014—about 27 times larger.[9] There is considerable debate on *when* the carbon energy age will come to an uneven and protracted close for most of the planet, but there is little debate on the fact that it will (cf. Mann 2013). The fact that oil and gas companies are now producing unconventional oil in difficult environments and at elevated prices and energy inputs is not a strong indication that the era can last a few hundred more years. This means that a post–carbon energy age will be historically inevitable. What this age might look like is an open question heavily debated by geologists, scholars and energy enthusiasts (e.g., Friedrichs 2010, 2013; Heinberg 2003; Kunstler 2005). To think about a post–carbon energy era, it may be useful to consider some of the transformations that have occurred from the age of efflorescences to the age of carbon energy (see table 2.0).[10]

Thus when we consider the nonrenewable nature of fossil fuels, it becomes possible to conceive of three major historical eras. Although the constitution and development of the post–carbon energy age is difficult to predict, I make some more detailed observations in the final chapter of this book. This does not mean that other divisions of time are useless or unneeded. But if we think on the scale of world history and the limits of the possible when it comes to energy supply and consumption, then conceiving of human political economies as existing in an age of efflorescences, an age of carbon energy and a post–carbon energy age seems justifiable. As stated earlier, this is not to suggest that there is any easy or straightforward way to demarcate these eras. Ultimately, this schema of time is a heuristic periodization rather than one that is concrete and true for each political community as a whole. The one thing that appears to be relatively clear, however, is that the increasing mag-

Table 2.0

Age of Efflorescences *60,000 years ago to 1850*	*Age of Carbon Energy* *1850 to 2150(?)*
Human habitats overwhelmingly in rural environments. By 1800 only 3 percent of the global population was urbanized (Di Muzio 2008).	Mass urbanization/suburbanization and the rise of thirty-three megacities with 10 million or more people. It is estimated by UN-Habitat that by about 2005 there were more people living in cities than in rural environments.
Local economies with long-distance trade carried out by a small fraction of a population. Save for luxuries for the rich, goods were largely produced and consumed within a ten-mile radius, particularly for inland communities (Scott 2011, 91).	The creation of a more integrated world economy with a global GDP of US$75 trillion in 2013 and US$23.4 trillion in world merchandise and services trade.[1]
Local rootedness, with the majority of humanity living and dying within an estimated ten-mile radius (Braudel 1983).	Globalization, automobilization and the rise of commercial air and cruise ship travel for the affluent and working classes of some countries.
Undeveloped transport networks with the fastest travel by sail and horses.	Annihilation of space by fuel.
For the majority, participating in the market is an opportunity rather than necessary for social reproduction and survival (Wood 2002).	For the majority of the population, the participation in markets is a compulsion for social reproduction and survival.
Variegated forms of human servitude from direct slavery to indentured labour.	Mass proletarianization and the rise of wage labour as the predominant form, though other forms of servitude remain.
The eventual formation of joint-stock companies to control various aspects of long-distance trade and European colonization.	The rise and proliferation of the modern capitalized corporation with vendible ownership titles and absentee owners.
Simple but changing division of labour.	Extreme specialization in the division of labour.
The slow rise of local credit-backed exchanges and sovereign control over coining money where gold and silver are deemed the only true international currencies among key rulers.	The rise of capitalist credit money no longer backed by a metallic substance but arguably backed by carbon energy. Mass monetization of socioeconomic space.
The slow rise of educational institutions and enrollment limited to a coterie of privileged students. Literacy and numeracy is low to nonexistent for most people throughout the period.	Mass and in many cases mandatory primary education typically sponsored by the public. Literacy and numeracy extended to 'the masses'. Democratization and proliferation of the university.
Variegated forms of political community with elite or privileged representation.	Greater democratization through polyarchic forms of democracy.

Age of Efflorescences 60,000 years ago to 1850	*Age of Carbon Energy 1850 to 2150(?)*
Superstitions and religions proliferate with over 2,500 gods invented.	Scientific revolution and the belief that all religion is human-made more widespread
Communication largely circumscribed by territory.	Mass worldwide and instantaneous communication is possible.
Slow population growth with the world reaching its first billion in 1804.	Exponential growth in population to 7.1 billion, up from 3.1 in 1962 and estimated to go as high as 9.6 billion by 2050.[2]
Wonderment at stars and the universe.	Space exploration is made possible and the composition of stars explained.

1 Global GDP data is from the World Bank and trade statistics from the World Trade Organization (http://data.worldbank.org/indicator/NY.GDP.MKTP.CD/countries?display=graph and http://www.wto.org/english/news_e/pres14_e/pr721_e.htm).

2 Https://www.un.org/en/development/desa/news/population/un-report-world-population-projected-to-reach-9-6-billion-by-2050.html.

nitude and universalization of the capitalist accumulation of money coincides with the age of carbon energy. Thus, in order to understand the relationship between energy, social reproduction and world order, we need a convincing theory of capitalism.

POLITICAL ECONOMY AND CAPITAL AS POWER

Plenty of scholars inside and outside IPE ostensibly study various aspects of capitalism, but few have ever asked themselves what the 'capital' in capitalism represents. Many are unaware that, from an analytical point of view, there are only two major schools of thought that try to provide us with an analytical definition of 'capital': the neoclassical and the Marxist (Nitzan and Bichler 2009; for summaries see Di Muzio 2014, 2015a). The neoclassicals took what could be called the 'Smithian turn' and built a theory of capital on the notion that it is a material and quantifiable entity. According to one popular text, capital 'is the set of tools that workers use: the construction worker's crane, the accountant's calculator, and this author's personal computer' (Mankiw 2009, 47). In this sense, capital represents a physical factor of production, such as a plant, machines or equipment, and though unsure how to measure it exactly, some economists even tack on 'technology'.

Although the word 'capital' has deeper historical roots (Braudel 1983, 232ff), in business parlance, before the *Wealth of Nations* (1776), capital meant a sum of money to invest or, alternatively, a sum of money already invested in expected profitable exploits. According to Cannan, however, Adam Smith redefined capital as material goods used in production. This was a 'very serious departure from the conception of capital which had hitherto prevailed. Instead of making the capital a sum of money which is to be invested, or which has been invested in certain things, *Smith makes it the things themselves'* (1921, 480; emphasis added). As the tradition of 'modern economics' sedimented in universities and textbooks years later, the tradition built upon Smith's initial confusion and made capital 'things themselves'. Prior to this point in history, capital was viewed as an auxiliary to production, not a primary input. The real 'factors of production' according to the classics were land and labour. But by theoretically making 'capital goods' a factor of production, the neoclassicals helped justify capitalist private profit and the societal distribution of income. As Nitzan and Bichler (2009) point out, this was largely the task of John Bates Clark in *The Distribution of Wealth* (1899).

Working with the idea that capital goods have their own unique productivity, Clark developed the infamous production function to demonstrate how each factor of production was rewarded according to its contribution to economic output. The only difficulty was demonstrating this empirically, and this is where the theory ran into unsolvable problems. The major problem is that to make this computation, we have to know the physical magnitude of labour, land and capital *before* we can determine output and therefore the distribution of income—that is, who deserves what based on their contribution. However, even if there was a clear and precise way to calculate the mathematically exact contribution from land and labour in modern finance, the value of capital goods is contingent on the income it generates. What this means is that as a factor of production, capital can have multiple magnitudes depending on what profit a company makes in the real world (for a more elaborate critique than allowed here, see Nitzan and Bichler 2009, chapter 5). Unable to come up with an independent magnitude measurable prior to uncovering the amount of profit, the production function breaks down and therefore neoclassical economics fails at one of its primary tasks: justifying unequal income to capital.

Karl Marx and his close followers have not fared much better. As already suggested, unlike the neoclassical framework, Marx's approach was both critical of class power and historically informed. He understood humans as productive social beings whose creativity was stifled and directed by capitalist owners in pursuit of profit and that the relationship between wage workers and capitalist owners grew out of feudal property relations. Marx was also deeply aware of some of the main contradictions of capitalism, such as the degradation of the industrial worker, the despoliation of the environment and the prevalence of underconsumption and surplus production when real demand was not backed by the ability to pay.

For all of Marx's many insights, when it came to explaining capitalist accumulation, he, too, chained his concept of capital to the rock of materialism. In contrast with the neoclassical 'factors of production' approach, Marx relied on the idea that the only source that could add value to nature to make commodities was human labour power. Marx argued that during the labour process, humans created more value than they were actually paid for during the length of the working day. As stated earlier, the problem was trying to demonstrate this scientifically—that is to say, to quantify the exploitation of the worker and show that the one and true source of capitalist profit was unpaid surplus labour. To do so, Marx and anyone convinced by his explanation for the origin of profit would have to somehow transform labour values or the quantity of labour contained in a commodity into prices. As already suggested, this has never been demonstrated, and as Nitzan and Bichler note (2009, chapter 7), there is also the further problem of making an analytical distinction between productive and unproductive workers in the Marxist framework. Since what is at stake in these analytical issues is not only mathematical precision, but also either the justification of capitalist profit (neoclassical) or the will to revolution (Marxist), the stakes are rather high. And since neither tradition seems capable of surmounting them (and likely never will due to their basic assumptions), we are left to consider an alternative theory of capitalism, not as a mode of production or some abstract space outside of history and politics but as a historical mode of power. By mode of power, we mean the specific architecture of power and logic that creates and re-creates a given hierarchical, class society where the emphasis is on scrutinizing and historicizing organized institutional power rather than *only* forms of production and labour practices.

Nitzan and Bichler suggest that one of the largest obstacles to overcoming our conceptual confusion is the general desire to interpret capital as something material. They propose that capital is not a material substance like machines, tools or equipment, but a form of differential social power that is commodified and owned by a small minority of the planet's inhabitants. The goal of the minority of owners and the firms they own and/or run is to generate differential earnings, thereby leading to differential levels of capitalization and political influence. Thus, the focus here is on what really matters to capitalists or investors—capitalization registered in a monetary unit. In the capital-as-power framework, accumulation, then, is rising capitalization, and differential accumulation is rising capitalization relative to a certain benchmark, like the S&P 500 or the MSCI World Index.

Although the precise computation of capitalization developed over time, today it can be understood as the process whereby a future flow of earnings is discounted into a present price and adjusted for some factor of risk. The computation is future oriented and relies on the time value theory of money in finance that suggests that a dollar or euro or any other unit of currency is worth more today than it is tomorrow. The primary reason why this is so is because a dollar, euro or other unit of currency can start earning interest right away. For this reason, the uncertain future is discounted. Thus, one of the key questions in the capital-as-power approach is to ask what is being capitalized when investors purchase claims on income-generating assets. The simple answer that any knowledgeable investor will be able to tell you is that you are capitalizing expected future earnings. But the capital-as-power perspective theorizes that the generation of earnings is not a simple matter of producing for the market, but a far broader power process that involves shaping and reshaping the global field of social reproduction and the wider conditions of existence. In essence, what is being capitalized is the corporation or state's power to affect social reproduction in such a way as to generate greater earnings faster than rivals. Those firms with the highest levels of capitalization are what Nitzan and Bichler (2009) call dominant capital, and their major owners are what I call dominant owners (Di Muzio 2015a). A brief example might help illustrate the approach.

According to the yearly *Financial Times Global 500*, a list of the 500 largest firms in the world by market capitalization, ExxonMobil is the second-largest firm in the world valued at US$422 billion. ExxonMobil is considered an oil and gas producer, and its income stream comes from selling crude oil, natural gas and various petroleum and chemical products to its customers. As an owner of ExxonMobil's shares, one of your key considerations would be its proven oil and gas reserves, which are booked as assets on the company's balance sheets. Increasing reserves means ExxonMobil has the ability to sell more in the future, and declining reserves means ExxonMobil has less to sell in the future. The former is a good indication for owners/investors, whereas the latter is not a strong signal of future earnings success—particularly if the company's reserves dip below its rivals'. Since ExxonMobil's income stream depends on being able to sell oil and natural gas and related products, it must constantly search out new reserves to replace or augment the oil and gas products it sells. But since earnings are a matter of power in the capital-as-power framework, ExxonMobil must do much more than simply find, produce and sell oil and natural gas products. As Steve Coll (2012) has documented, the company can be considered a 'private empire' that has seen governments come and go over the course of its 100-or-so-year history. During this time the company has exerted its organized corporate power to generate record earnings and capitalization. Although we cannot trace all of the company's attempts to exert power over the social process, consider some of the following actions that all entered into the company's differential capitalization since 1999:

- Lobbying the European Commission, the US Federal Trade Commission and politicians on Capitol Hill in order to get its merger with Mobil approved in 1999
- Lobbying the US State Department to lift sanctions on Libya to exploit its oil wealth
- Volunteering to train Iraqi oil workers
- Influencing public perceptions by funding institutions who deny global warming and advising the George W. Bush administration to abandon the Kyoto global warming treaty
- Hiring and supplying local military personnel in Aceh in order to protect its natural gas fields
- Garnering public funding from the World Bank in order to help finance its operations in Chad and Cameroon
- Potentially conspiring with British Petroleum to restrict the supply of natural gas from Alaska
- Appealing the US$4.5 billion settlement awarded in a class-action suit to victims of the *Exxon Valdez* oil spill
- Influencing US Vice-President Cheney's Energy Task Force during the first George W. Bush administration
- Lobbying to drill in Alaska's Arctic National Wildlife Refuge (the previous examples are from Di Muzio 2007, 522–23).
- Sabotaging oil wells in Texas by pouring them in with cement, explosives, sludge and garbage to prohibit other firms from exploiting the wells[11]
- Encouraging the revolution in fracking while CEO Rex Tillerson went to court to stop fracking in his own backyard[12]

The list is certainly not exhaustive, but the point is that all of these attempts to exert power over the social process affected the company's earnings and therefore its capitalization. Production does indeed matter, but the factors that contribute to earnings and capitalization are far broader than 'worker exploitation' or the value of 'capital goods'. Indeed, instability in the Middle East, if considerable, can cause oil prices to soar and can boost Exxon's profits and capitalization, as it did throughout the war on Afghanistan and the invasion of Iraq. In this way, the emergent capital-as-power perspective within critical IPE is a much more realistic and convincing theory than its Marxist and neoclassical alternatives. For those readers who are encountering this theory of capital for the first time, it may be best to summarize its basic assumptions and postulates in a clear and systematic way, given that the perspective informs the bulk of this study:

- Although the capitalist universe is replete with companies, those with the largest capitalization—what Nitzan and Bichler call dominant capital—are the most powerful actors in shaping and reshaping global forms of social reproduction. By far, the two most capitalized sectors of the global economy are oil and gas (*control of energy*) and banks (*control of the money supply through loans*).
- Through its national debt, the state can also be considered a capitalized entity since its debt is owned by private social forces. There is no separation, practical or theoretical, between 'politics' and the 'economy'—they are fused.
- Investors do not capitalize profit per se, but expected future earnings adjusted for some factor of risk.
- Earnings, and government fines and taxes, are not simply a matter of production but result from an entire range of power processes exerted by corporations and states over social reproduction.
- Companies do not aim to maximize profits but to accumulate differentially and use benchmarks to help assess their performance relative to their rivals'.
- Accumulation is understood as rising capitalization measured in pecuniary units.
- Those who own significant claims on dominant capital can be considered dominant owners.
- Since earnings and capitalization are a matter of power, capital is theorized as a mode of power rather than a mode of production connected to industrialization.
- Capital is not a material thing but the commodification of differential power symbolically represented in money units.

In the final section of this chapter, I illustrate how a focus on what I call the capitalization–energy–social reproduction nexus can help us explain and understand transformations in the global political economy of capital as power.

THE CAPITALIZATION–ENERGY–SOCIAL REPRODUCTION NEXUS

In this work, I propose, as I have done so in a previous work (Di Muzio 2014), that one of the most interesting ways to consider the emergence and development of carbon capitalism and its related petro-market civilizational order is to focus on the interconnections between capitalization, energy and social reproduction. The reasons for this focus are fairly straightforward. First, capitalization is the dominant ritual of modern capitalism. Although there may be 'varieties of capitalism' operating in local cultural registers and regulatory frameworks, capitalization remains the all-encompassing ritual of capitalists just as much as they pursue the logic of differential accumulation (Hall and Soskice 2001). For example, regardless of the variety of practices, we would be hard pressed to find capitalists seeking to deaccumulate, chief executive officers chasing less profit each quarter and the absence of important consensual benchmarks agreed to by the global finance industry. What this suggests is that if we want to find the origins of capitalism, we should be looking for early forms of ownership and capitalization and how these early forms emerged and were organized into markets for securities. According to Michie, by 1910 the value of all outstanding securities was estimated at £32.6 billion, or in inflation-adjusted dollars in 2014, US$5.5 trillion (2006, 6). These securities were owned by just 1 percent of the global population at the time (Di Muzio 2014, 23).

According to McKinsey and the World Federation of Exchanges, today's outstanding capitalization is roughly US$225 trillion, or an increase in debt and equity from 1910 of 4300 percent. Thus, within the span of just over a century (1910–2014), global capitalization has ballooned to epic proportions. Since these securities capitalize expected *future* earnings, this means that global expectations have also increased. What could account for this stupendous rise, particularly since the latter half of the century, as figure 2.1 reveals?

Although it is certainly not the only factor, given our knowledge that energy is the capacity to do work and surplus energy the capacity to do greater work, then we might suspect that the rise in capitalization to correspond with increasing energy consumption across the world. Using data from the EIA,

Table 2.1

Debt and Equity	*1990 US $trillion*	*2012 US $trillion*	*Percent Increase*
Government Bonds	9	47	422%
Financial Bonds	8	42	425%
Corporate Bonds	3	11	267%
Securitized Loans	2	13	550%
Nonsecuritized Loans	23	62	170%
Equity	11	50[1]	509%

1 World Federation of Exchanges as recorded in September 2014 that equity stood at US$67 trillion.

(McKinsey Global Institute 2013: 2)

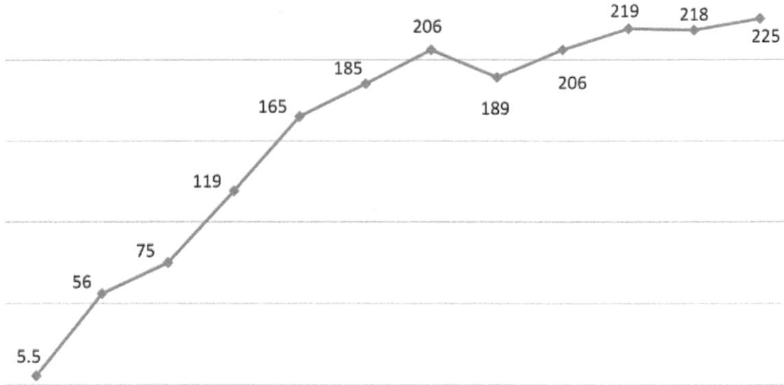

Figure 2.1. Global Financial Assets: Total Capitalization (US$ trillion). (McKinsey Global Institute 2013

figure 2.2 considers global fossil fuel consumption since the 1980s, and figure 2.3 looks at the world's total final energy consumption.

With statistics for the world stretching back to 1980, the story here is rather straightforward. Energy consumption for all three fossil fuels has steadily increased since 1980, and total final energy consumption measured in British thermal units (BTUs) increased by 84 percent for the period. However, this revolution stretches back to the first mass and sustained use of fossil fuels

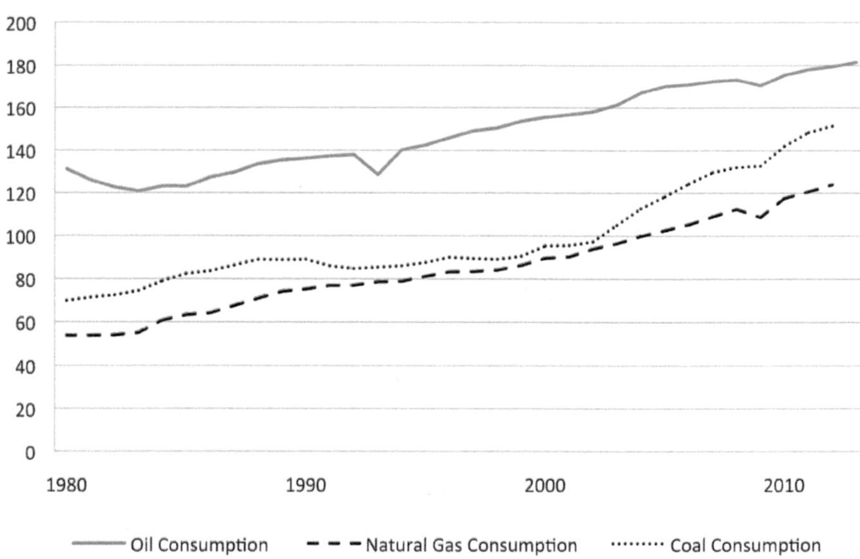

Figure 2.2. Total Fossil Fuel Consumption (Quadrillion btu)

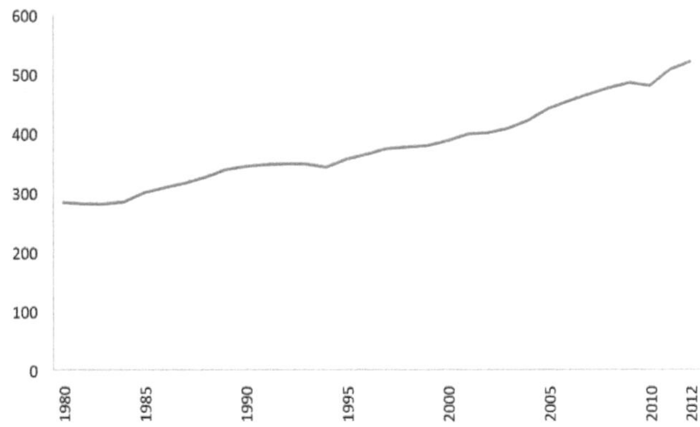

Figure 2.3. World Total Final Energy Consumption (Quadrillion btu)

in Britain. Smil estimates that pre-agricultural societies consumed about 9.5 million BTUs a year, whereas by the end of the 1800s, Britain was consuming 95 million BTUs per capita. In 2005, the United States, the largest national economy on the planet, was consuming 313 million BTUs per capita. And as figure 2.2 reveals, the world consumed about 520 *quadrillion* BTUs in 2012, or roughly 55 *billion* times more energy than pre-agricultural societies.[13] In this way we can start to think about how the increasing magnitude of capitalization over time corresponded with a revolution in fossil fuel consumption. Smil summarizes the great transformation:

> Traditional societies were thus, at least in theory, energetically sustainable on a civilizational timescale of 10^3 years, though in practice many of them caused excessive deforestation and soil erosion and overtaxed their labor. In contrast, modern civilization rests on indubitably unsustainable harnessing of a unique solar inheritance that cannot be replenished on the civilizational timescale. This dependence has given us access to energy resources that, unlike solar radiation, are both highly concentrated and easy to store and that can be used at steadily higher average rates. *Reliance on fossil fuels has removed the limit that the inherently low photosynthetic efficiency and low-level conversions of animate, water, and wind energies imposed on human energy consumption.* As a result, the total energy flux through civilization has risen steadily to unprecedented levels. (In Chiao et al. 2010, 712; emphasis added)

In other words, the transition to consuming ever greater amounts of fossil fuels has given us the greater, but radically uneven, capacity to do work on the natural world and our own built environments, while at the same time oil and gas corporations seek out even more energy to continue patterns of globalized social reproduction that are highly dependent on carbon energy, leading to the carbonization of everyday life. It is my contention that private ownership and the power of the firm were decisive moves to capture and control energy stores and flows for private benefit. At first, the scale was meager, with the 'national' debt and a few joint-stock companies capitalized on nascent securities markets. For example, as late as 1840, the public debts of Britain and foreign governments on the London market made up 89 percent of all traded securities.

However, as Michie's study documents, from 1850–1900, the number of listed companies and the globalization of stock exchanges started to soar, as did capitalization and public debts of European states engaged in international colonialism and warfare. From a handful of major transnational companies, by 2010 there were 45,508 listed on the major exchanges of the world, up from 20,726 in 1990.[14] The top 500 global corporations by market capitalization make up about 42 percent of global market capitalization at US$31 trillion of the US$67 trillion outstanding.

Thus, capitalization and the institutions and regulations to support it have grown massively since the more thoroughgoing exploitation of coal, oil and natural gas. But since capitalists or investors capitalize expected future earnings, and earnings, I argue, are a matter of exerting power over the social process, we need to consider capitalization, energy *and* social reproduction as practically and theoretically interlinked. As I have argued earlier, social reproduction in the carbon era is far more globalized than it was during the era of efflorescences, and we can use the proxy of world trade to get an idea of its historical novelty and its scope. Since 1950, when statistics are available, global merchandise and commercial services trade increased by 4580 percent from US$500 billion to US$23.4 trillion in 2013.[15] Although global trade is only a proxy for what I have called globalized social reproduction, it should be rather clear that this amplification in trade has coincided with oil as the main fuel of transport and mass containerization. In fact, the first containers were used in 1956 and went from Newark to Houston on a refitted oil tanker. As Levinson explains:

> The container is at the core of a highly automated system for moving goods from anywhere, to anywhere, with a minimum of cost and complication on the way. The container made shipping cheap, and by doing so changed the shape of the world economy. This new economic geography allowed firms whose ambitions had been purely domestic to become international companies, exporting their products almost as effortlessly as selling them nearby. In 1956, the world was full of small manufacturers selling locally; by the end of the twentieth century, purely local markets for goods of any sort were few and far between. (2006, 2–3)

Although this statement may be a bit exaggerated, today, about 90,000 cargo tankers travel the ports of the world dispensing goods and creating billions of dollars in health costs due to the pollution emitted in the course of business.[16] In a way, then, modern carbon capitalism is also container capitalism.[17] What this discussion on globalized social reproduction suggests is that before the widespread use of carbon energy, transnational market forces did not mediate or arbitrate the everyday lives and lifestyles of the majority of the world's population. The transition to the current global social order where the allocation of goods, services and life chances are mediated and arbitrated by transnational market forces and mechanisms has only been a few centuries in the making—accelerating and amplifying mostly after World War II if we are generous. Modern forms of social reproduction and survival are now wholly dependent upon having access to goods and services for purchase on the market. One glaring indication of this transition has been the demographic shift away from a world society where the majority of the population was primarily rural to one that is increasingly urban, suburban and, for over 1 billion people, living in inadequate shelter. By one estimate, 79 percent of the global population will live in cities by 2050.[18] With little to no access to land and the means of production outside of wage relationships, the vast majority of city dwellers are completely dependent on markets, whether they are used to gain employment, shelter, clothes or food. Furthermore, the logistics of global supply chains have expanded so far geographically that whole populations are virtually 100 percent reliant on the vast networks of supertankers, cargo planes and diesel trucks moving not just computers and toys but essentials like food, fertilizer and medical equipment. They are also reliant on a vast energy-intensive infrastructural network of ports, freight terminals, loading docks and airports (Rodrigue et al. 2006). In other words, only a fraction of the global population could be considered even remotely *fully* self-sufficient in terms of fulfilling their basic needs like a proper diet.

In sum, then, rather than consider energy, capitalization and social reproduction as distinct categories, I argue that they should be understood as deeply interconnected. Only then can they be mobilized to help us explain the rise and fall of carbon capitalism and petro-market civilization. In the following chapter we will consider the rise of petro-market civilization in Britain.

NOTES

1. In the foreword to their seminal book, Chilingarian and Yen (1978) write the following: "As the fact of depletion of the major fossil-fuel sources is gradually being unveiled on a worldwide scale, the exploration and exploitation of the earth's other natural fuels assume ever-increasing importance. Because of practically global

availability of bitumens, asphalts, and tar sands, these natural deposits could offer a significant supplement to the waning oil and gas reserves".

2. Originally published in French in 1986.

3. This largely ignores what land was used for and how it was unevenly distributed.

4. Linebaugh (1976) argues that Marx first realized the importance of political economy when, as a journalist, he was confronted with the mass criminalization of the Mosel peasantry for collecting wood on 'private property'.

5. Huber (2008) argues that historical materialism can be 'energized'. Although this article is brilliant, in my opinion it remains tethered to Marx's mistaken labour theory of value and therefore has theoretical and practical implications beyond Marx's analytical approach to accumulation and prices.

6. See also the literature cited in chapter 1.

7. I cite here major examples of the literature that have a political economy focus, or at least deal with questions related to energy and the economy. The more security-oriented literature has been cited earlier. This list of citations also ignores important contributions from specialist journals like *Energy Policy* and *Energy* and is meant to be representative rather than exhaustive.

8. http://www.iea.org/topics/coal/.

9. Data are from the Global FT 500, 2014 and The Stowe Global Coal Index.

10. This comparative list is meant to be illustrative, not exhaustive.

11. http://www.bloomberg.com/apps/news?pid=newsarchive&sid=aH4MoH2m4Z0w

12. http://www.forbes.com/sites/rickungar/2014/02/22/exxon-ceo-profits-huge-as-americas-largest-natural-gas-producer-but-frack-it-in-his-own-backyard-and-he-sues/

13. Quadrillion = 1,000,000,000,000,000

14. Data are from the World Federation of Exchanges (http://www.world-exchanges.org/statistics/time-series/number-listed-companies)

15. Current data are from the World Trade Organization (http://www.wto.org/english/news_e/pres14_e/pr721_e.htm).

16. http://www.theguardian.com/environment/2009/apr/09/shipping-pollution.

17. The BBC tracked a container for over a year and reported that in total mileage, it did over two laps of the earth (http://news.bbc.co.uk/2/hi/business/8314116.stm).

18. John Vidal, 'UN Report: World's Biggest Cities Merging into 'Mega-Regions', *The Guardian*, 23 March 2010.

3

The Birth of Petro-Market Civilization in Britain

Given their understanding that humans cannot be separated from the natural world, if Marx and Engels were alive today, it is highly likely that they would be keenly aware of climate change and the importance of fossil fuels to the social reproduction of high-energy–intensive modes of life and geopolitics. Yet when they were writing, they marvelled more at social labour than they did the energy source that completely underwrote its development:

> The bourgeoisie, during its rule of scarce one hundred years, has created more massive and more colossal productive forces than have all preceding generations together. Subjection of nature's forces to man, machinery, application of chemistry to industry and agriculture, steam navigation, railways, electric telegraphs, clearing of whole continents for cultivation, canalization or rivers, whole populations conjured out of the ground—what earlier century had even a presentiment that such productive forces slumbered in the lap of social labor? (Marx and Engels 1848, 7).

It may seem strange, then, that Marx treated the most renowned and used fossil fuel of his time—coal—as an auxiliary in production rather than a decisive factor in facilitating the 'more massive and more colossal productive forces' than all preceding generations combined (Alam 2009, 171; Georgescu-Roegen 1971, 1976; Marx 1887; see also Malm 2013, 50).[1] Instead, the locus of this massive transformation was pinpointed to exploited social labour with the magnitude of capital accumulation tightly chained to the labour theory of value. But forms of social labour existed long before capitalism, the rise of stock market capitalization and the modern multinational corporation. Following Mumford (1934), we can understand the building of the monumental architecture of the Egyptian pyramids and the Roman Empire as two prominent examples of social labour. Both required a system of quantification, directed labour and extraordinary planning and coordination. Yet the hierarchical power civilizations of Egypt and Rome did not achieve the 'colossal productive forces' on the scale noticed by Marx and Engels in industrializing Britain. Historians and political economists have long pondered why it was that the economically backward island of Britain was the first to industrialize on such an incredible scale, leading to the largest global formal empire on the planet and an observable divergence of wealth between it and the rest of the world by the nineteenth century. British international supremacy effectively ended with the ruling-class–financed and orchestrated carnage of World Wars I and II. In this chapter we will consider some of the reasons given to explain this divergence, but instead of focusing on what could be called the 'rise of the West' literature, I will focus on how political economists have theorized the origins of capitalism in England.[2] The argument in this chapter is that the prevailing accounts, although incredibly insightful in many respects, have an inadequate and often misleading account of the rise of capitalism. This is so for at least two reasons. First, most 'transition to capitalism' scholars, like Marx, treat energy as auxiliary or ignore it altogether. In this formulation, capitalism is either born with merchants in the city waiting to invest their money gained in trade into factory production, or with capitalist tenant farmers in the countryside increasing productivity and exploiting rural wage labourers. Second, as Anievas and Nisancioglu (2014) rightfully point out, the prevalent view of political Marxists is too conceptually narrow and tunnels empirical research in space and time by locating capitalism *solely* in the English countryside where it eventually diffuses outward.[3]

My own view advanced in this chapter is not to explain the origins of capitalism per se, but to focus more on the factors that contributed to *an amplification in orders of magnitude of capital accumulation and its accelerated sociospatial universalization.* Capitalism, I will argue, does not first appear in England—as even Marx and later Braudel recognized. What is special about England is that a confluence of events—one may want to call them a confluence of revolutions—related to energy, social reproduction and capitalization set in motion the capacity for an ever-greater accumulation of money and accelerated sociospatial expansion of mechanization and dispossession. After a brief critique of the 'origin of capitalism' literature, I elaborate on my own argument in the remainder of this chapter. First, I consider the limits of capitalization during what I have called, following Goldstone, the age of efflorescences. I then turn to the confluence of revolutions that contributed to the intensification (increasing orders of magnitude) and extensification (sociospatial scope) of capitalism during the opening stages of the age of carbon energy. The chapter concludes with a consideration that imperial Britain could be conceived of as the world's first empire of coal and how Britain's ruling class eventually transitioned to a petro-market civilization more reliant on oil, chiefly but not exclusively, for military purposes.

THE ORIGINS OF CAPITALISM

Although perhaps marginal to the field—particularly the ahistoric open economy politics approach—debates on the origin of or transition to capitalism have been perennial in political economy scholarship (Anievas and Nisancioglu 2013; Beaud 2001; Bhambra 2010; Braudel 1983; Brenner 1976, 1977, 1978; Byers 1996; Dimmock 2014; Dobb 1946/1963; Harman 2004, 2006; Heller 2011; Hilton 1976a, 1976b; Hobson 2004; Lefebvre 1982; Mielants 2007; Nitzan and Bichler 2009; Sweezy 1954; Tawney 1926; Wallerstein 1974; Wood 2002; Zmolek 2013). The debates are extensive and too lengthy to approach in microscopic detail here, but a reasonable summary of the main positions in the debate can be summarized with a view to critiquing them and providing an original theory of what I have called carbon capitalism. At issue in the transition debate is essentially how and why the so-called feudal mode of production was gradually overtaken by the capitalist mode of production, or in some formulations why England and Europe industrialized first and other countries failed to go through a synchronic industrial revolution. In this latter formulation, the Industrial Revolution is largely conflated with capitalism, and in many senses industrialization is interpreted as an historical development that is desirable for societies and one that noncapitalist populations ought to pass through on their way to communism or mass consumption. This is largely in keeping with Marx and Engels' only very partially correct notion that industrial-capitalist countries show what the future holds for the nonindustrial and noncapitalist societies.

Another feature of the transition debates is the attempt to find primary or, in some cases, singular, causes for the transition to capitalism internal to feudal agrarian relations or external relations related to the world market. Marxist scholars give various weights of importance to these factors while downplaying others deemed less crucial to the transition between two different modes of production. As I hope to show, if only in brief, there are many very peculiar things about these debates—peculiar because they have largely been ignored or thought auxiliary. First, most scholars in the debate never clearly define what 'capital' or 'capitalism' is and therefore there is considerable confusion over exactly what we should be looking for historically when we try to identify the origins or birth of capitalism: a transition in social property relations, the birth of ownership, the accumulation of money, the Industrial Revolution, the production of commodities, the increase in productivity per worker, financial innovation, the development of the forces of production, international trade, colonization, a complex division of labour, the specialization of industry, the creation of wage labour, the abolition of slavery, the birth of the world market, the centralization of state violence and demilitarization of the nobility, an accumulation of investable merchant capital, and the list could continue. As one reviewer noted, '[I]n arguments about the origins of anything, definitions become, or should be, paramount' (Lawrence 2008, 503). Yet even where definitional issues are sorted out by calling capitalism a 'mode of production' that produces commodities for profit and where workers add more value to the commodities produced than they are paid throughout the working day, we still run into considerable theoretical and practical problems. Although there is no doubt that Marx considered capitalism a 'mode of production' and a social property relation between wage workers and owners of the means of production, when it came down to his analysis of profit and accumulation, he argued that 'capital is dead labour, that, vampire-like, only lives by sucking living labour, and lives the more, the more labour it sucks. The time during which the labourer works, is the time during which the capitalist consumes the labour-power he has purchased of him' (Marx 1887/1996, 160). It is perhaps small wonder then that modern Marxists are all over the map on what to look for when they go in search of the origins of *capital*ism. If they took Marx at his analytical word, they *should be* looking for used-up or expended labour during a paid working day (the definition of dead labour). This, of course, would be absurd since we can find waged-labour–producing goods in different 'mode of production' eras. Problems only compound from here. The core of Marxism and the raison d'être for revolution and the search for a new society founded on communal ownership and democratic planning is anchored in the labour theory of value. But as discussed in chapter 2, Marxists have been unable to show how labour time values can be *transformed* into prices during the production of commodities. No one should doubt that capitalists employ human labour power/energy throughout a working day, but it does not follow that this form of employment *alone* can explain prices and the magnitude and distribution of profits and capitalization. Furthermore, Marxists make an untenable and ultimately subjective division between productive and unproductive workers (Nitzan and Bichler 2009, chapter 7).

Second, other than definitional issues, the transformation problem, and sorting out the distinction between workers who create surplus value and those who merely consume it, there are at least three additional oversights in the debates. First, with some minor exceptions already noted, there is no general recognition of the importance of energy or the fact that energy scarcity may have played a decisive role in shifting England on to a different path in major contributions to the political economy debates (cf. Huber 2008). Second, the capitalization of the English state and the decisive role played by the Bank of England is also generally ignored as a key factor in explaining the origin and development of capitalism. Why this is so is very peculiar, given that debates leading up to the Bank of England argued that although the economy could be more 'productive', there was always a dearth of money in the economy to propel 'improvements' or what we would today call 'economic growth'. As Wennerlind observed, '[W]hile modern economic theory does not recognize the possibility of a scarcity of money, seventeenth-century thinkers were *consumed* by this problem' (2011, 17, my emphasis; see also Di Muzio and Robbins 2015). Since you cannot have Marxist or liberal 'endless accumulation of capital' without an expanding money supply, it seems strange that there is virtually no discussion on the necessity of credit and the creation of the Bank of England as a definitive moment in the constitution of capitalism. Money and credit, as it were, appear as if by magic in the transition debates: we are accumulating it, but we know not from where it came! Third, the majority of the transition debates either totally ignore or severely underplay the fact that Britain's ruling class, with few exceptions, was at virtually constant warfare since the Norman Conquest of 1066. If, as Tilly and other sociologists have suggested, war makes the state and the state makes war, then it seems strange not to view the need for war materiel as a fundamental driver of capitalist extraction, production and accumulation (Evans et al. 185, 169–91). This is all the stranger given the fact that warfare and the preparation for war continue to be constant features of modern capitalism. The capitalization of the aerospace and defence sector is a half-trillion dollars, making it the twenty-fifth largest sector of the global economy (out of forty-two).[4] Why should we find the origins of capitalism *solely* in the rural countryside with relatively pacific landlords interested in making improvements to agricultural production because they supposedly could not inflict violence upon their subordinates? Why shouldn't we look to perpetual war and geopolitical competition as key drivers of social change in an era when there were clear limits to accumulation due to limited energy use? Isn't this, at base, the essence of capitalism: constant war for differential accumulation, for social status, for power, for the extension of ruling-class power? As we will discuss more in detail, slavery and colonialism are also considerable omissions in the debates.

Viewing capitalism as a mode of power rather than a mode of production, my approach to explaining and understanding the emergence of capitalism and its increasing magnitude and spatial scope is to start with a convincing definition free of the materialism of Marx's dead labour and the liberal's capital goods. As Nitzan and Bichler have convincingly argued, capital can be theorized as *commodified differential power* measured in a monetary unit. In fact, they go so far as to suggest that the only reason capital can be accumulated at all is because it is not a physical thing or material entity (2009, 271). In other words, no capitalist pursues the collection of more 'dead labour' in the form of commodities for sale or more of Mankiw's capital goods *as an end in itself*. If they are not doing this (and clearly they are not), then we need a far more convincing explanation than Marx and Marxism, as well as neoclassical economics, have so far allowed. Working with the idea that capital is not a physical entity but the simultaneous commodification and quantification of a social power process means that any qualitative power process can *potentially* be commodified, priced, owned, or traded. The value of capital is not determined by production alone. Indeed, as we will encounter many times throughout this study, capitalists must be vigilant to sabotage production as a going concern. This sabotage enters into the ritual of capitalization when expected future earnings are capitalized. Capitalization, or the process whereby expected future profit is discounted into a present value, stretches back to the Italian city-states of the thirteenth century and likely before. But according to Nitzan and Bichler: "the first *systematic rules* of discounting were laid down already in the mid-nineteenth century, by a group of German foresters (Faustmann 1849). The foresters tried to figure out how they should value wooded land and the associated activities of planting and harvesting, and in the process developed many of the mathematical formulae of present value" (2009, 155–56, emphasis added).

Perhaps not surprisingly given Marx's concern for history and deriving theory from actual social relations, there are passages where he clearly understood the capitalization process. Although capitalization is mentioned only once in *Capital* volume 1 and there are attempts to deploy the concept in *Capital* volume 2, it is only in volume 3 where Marx discusses capitalization at any length. Marx appears to realize that modern capitalist ownership is vendible and that capitalists are trading claims to expected future profits generated by businesses. He also notes how the value of these claims fluctuates depending on the rate of interest (the normal rate of return) and whether earnings expectations are in fact realized.[5]

The problem is that rather than affirm that this is indeed what capitalists are doing at the upper echelons of the capitalist hierarchy and readjust his theory, he ignores the problem and writes these claims off as 'fictitious capital' (Marx 1894/1996). The idea of 'fictitiousness' is interesting given our interpretation of capital as a nonphysical entity. But the similarity is only superficial. By fictitious, Marx appears to have meant 'unreal' or 'illusory'. He suggests that ultimately these claims and their shifting valuations do not really matter. What appear to matter for Marx are not ownership claims, but the real physical entities they capitalize and the workers using them to generate more profit. At one point he gives the example of a railway company and argues that even if the valuation of its shares become worthless, the real wealth of the rails and railway cars are still present and can be used; real wealth has not diminished (Marx 1894/1996: see chapters 29, 37 and 38). In some senses, this is a valid point—the physical machines and infrastructure would still be there—but that is not what capitalists are concerned with as an end goal. What matters to them are the future returns the railway is expected to generate relative to competitors and the return on government securities. If the securities have no value, like those of Enron, we can assume that the business concern had either no earnings or no significant differential profit, and that is the only thing that matters to capitalists. Marx may have thought this an absurd situation, but it does not make it any less real for owners or their money managers looking into the future and trading ownership claims to earnings. There is zero doubt that the expectations of capitalists may be completely wrong or never materialize, but that does not stop them from looking to the future and capitalizing income-generating entities as though expectations will be met based on the available information. The act of capitalization is the dominant ritual of capitalism, and this is why Nitzan and Bichler argue that capital is commodified differential power, not dead labour and not equipment, machines, or technology. The failure to understand this among scholars and the general public is akin to an astrophysicist thinking that the sun is still made of iron after the groundbreaking work of Cecilia Payne in astrophysics.

If it is accurate to claim that capital can be theorized as commodified differential power and that anything that can generate an income stream can potentially be owned and capitalized for profit, then we have a different starting point than the current origins of the capitalism debate in political and other forms of Marxism, not to mention the rise of the West/great divergence literature. It means that we can find the 'origins' of capitalism anywhere social power is owned, commodified and capitalized. What I would argue matters most is not to pinpoint a precise origin with an exact date and time for the emergence of capitalism, but to focus on the factors that contributed to an amplification in the magnitude and universalization of accumulation, which always leads to qualitative transformations in social life. Based on this recognition, my theoretical argument will be sketched next, and the remainder of this chapter uses the energy–capitalization–social reproduction nexus to explain why the intersocietal relations of England were so special when it comes, not to understanding the origins of capitalism per se, but its growing magnitude and universalization.

First, significant differential power that can shape and reshape forms of social reproduction appears historically in organized, institutional or corporate form. The three most important have been organized government with the power to tax, spend and enact laws with the control over the overwhelming means of violence; any corporate force generating an income; and mercenary armies. Second and crucially, where metallic money was in circulation, sovereigns typically had the power to coin money, and—at least for the period we are concerned with—authoritative or dominant money that could also be used internationally was primarily understood to be the metallic substances of gold and silver. Thus the amount of dominant money was limited by the supply of gold and silver and to get more of it meant doing one or all of the following: (1) finding and exploiting a new mine, (2) plundering it or taxing it away from others and/or (3) trading commodities on the market for gold and silver or paper claims redeemable for gold and silver. Forms of credit existed and could be geographically extensive, but these relationships were largely private and personal between known lenders and borrowers. Goldsmiths did loan in excess of their gold deposits, but these credit notes were neither generally assignable nor transferable and therefore could not act as a circulating national currency or expand the money supply in any significant way (Davies 2002, 249ff; Wennerlind 2011, 69).

The real innovation, as I will discuss in greater detail, was the subordination of the Crown to Parliament in the seventeenth century and the eventual creation of the Bank of England (1694). The English Civil War (1642–1651) and the Glorious Revolution (1688) juridically forged the modern institution of private ownership and placed the power to tax and spend in parliamentary rather than sovereign hands. Up until this point, the tenure of property in land was a grant or privilege bestowed by royal authority—one that could be (and was) infringed upon by arbitrary taxation or confiscation (Commons 1959, 217; Levy 1983, 119).[6] Loans, too, were often forced from subjects in control of substantial sums of coinage. The major aim of the Parliamentarians in the English Civil War was to overthrow this arbitrary relation of royal power and to install their own mode of power premised on ownership, the accumulation of money and the control of the state by crushing more democratic projects (Hill 1985; Kennedy 2006, 2008). In this sense, the major social relation we should be concerned with is not between a class of capitalists and wage labourers, but between capitalist owners and nonowners.

The Bank of England is also a crucial development because its initial loan to Parliament represents the first major *private* and *permanent* capitalization of the state's organized differential power over the population of England and its colonies. The loan, as in subsequent cases, was for interdynastic ruling-class war—in this case the War of the League of Augsburg (Davies 2002, 259). The permanent loan also served to expand the money supply, allowing for greater magnitudes of money to be accumulated or invested in income-generating enterprises. Third, the money supply expanded through interest-bearing debt over time because the exploitation of coal (combined with other energy sources) gave the ruling class of owners a greater and geopolitically differential capacity to generate earnings across a range of fields, from ironworks to ship making and from locomotives to war materiel. War and the preparation for war were absolutely crucial and required constant energy. But not only did constant warfare or its threat require constant energy, it also contributed to significant 'spin-off' technologies that could be privately capitalized—a trend that continues to this day from commercial aviation all the way to drone technology (Trebilcock 1969).

The exploitation of coal related to state power and war was not the only factor that contributed to the growing magnitude of capital accumulation and its greater universalization, but it was absolutely essential and historically exceptional when we consider the magnitude of consumption in the British case (Wrigley 2010). This period also corresponds with a revolution in scientific knowledge made possible by carbon energy (Nikiforuk 2012, 158ff). Fourth, the expropriation of direct producers from the land and the loss of customary rights and access to common fields did not immediately create a pool of useful wage labourers for capitalists as some suggest. As both Marx (1887) and Polanyi (1957) remind us, the enclosure movement and additional measures used to dislocate people from the land, such as restrictive forest and hunting laws, created mass pauperization, not mass proletarianization (Perelman 2000).

What to do with the landless poor became one of the chief problematics of early political economists, but while these debates raged, in practice, mass pauperization was met with a mixture of punitive measures and parish charity depending on the time, place and case. What this suggests is that capitalist owners did not require a mass pool of wage labourers until quite late; that is, in the nineteenth century when the combustion of coal gave the ruling class more means to shape and reshape industrial social reproduction. Indeed, Polanyi (1957, 101) dates the birth of the English working class to 1832 when the meagre welfare system that at least provided some social protection to the unwaged, underemployed and landless was repealed (see also Denning 2010). If the unemployed or their ancestors had been dislocated from the *means* of subsistence, now they were dislocated from subsistence itself and left to the vagaries of the labour market. Last, since capitalization and vendible ownership claims are key features of the capitalist mode of power, one crucial dimension of explaining the magnitude and universalization of capitalization is the presence of regulated stock exchanges for raising funds, price discovery for new companies and organizing and reorganizing ownership claims. In England, the official history of trading claims to assets goes back to 1571 with the creation of the Royal Exchange in London. But its current incarnation, the London Stock Exchange, was not inaugurated as a regulated exchange until 1801. It is important to point out that even by 1840,

89 percent of the securities traded on the exchange were claims to the public debts of governments in Britain and abroad. 'In other words, the largest financial game in the world was the capitalization of the state's power to tax its citizens' to service the claims of wealthy bondholders—the effective owners of these nondemocratic governments (Di Muzio 2014, 25). In this sense, the state can be conceived of as the first 'modern' corporation whose democratization for nonowners only happened gradually and with a process of social struggles, from workers to women and the racially subordinated. It is only comparatively late when the capitalization of modern corporations outpaced the magnitude of public debts around the world—likely at some point in the 1980s (McKinsey 2013, 2).[7]

In November 2014, the capitalization of the London Stock Exchange stood at US$3.3 trillion, only surpassing the trillion-dollar mark in 1986. The public debt of Great Britain as of the same time is US$2.6 trillion or about 1.7 trillion pounds sterling. Thus in the grand scheme of human history, the organized political power of owners in government, an expanding money supply premised on the capitalization of the state's power to tax, surplus carbon energy, a population no longer chained to the soil in the majority, and the capitalization of joint-stock companies on stock markets come together in a relatively short period in history—first in Britain and later elsewhere. The burden of my argument is to demonstrate this historically, and it is to this subject that I now turn.

THE AGE OF EFFLORESCENCE: THE LIMITS OF CAPITALIZATION

According to Goldstone, there is a growing consensus among historians that social reproduction in the early modern era should not be characterized as inert or static as some earlier historians argued. Rather, our evidence now suggests that earlier societies should be theorized in movement, fishing and working the land in various ways, innovating according to ability, environmental conditions/endowments and need within developing cultures of meaning. There is little doubt that the majority of the population was primarily involved in subsistence-related work and land was the primary source of wealth. All known agrarian political communities were hierarchically arranged so that a small minority at the apex of society received the majority of the benefits and lived substantially better than those they were able to subordinate by force, mysticism, or some degree of consent. During this era, having considerably more than someone else largely meant taking it away from others, whether through direct violence/appropriation or by confiscating it through law/custom backed by an established authority and the threat of punishment.

After the breakdown of the Roman Empire (27 BC–476 AD), most of Europe consisted of various centres of royal or princely authority achieved through force, heritage and/or intermarriage. The Holy See also extended its power outward and had considerable authority over interdynastic relations until the Protestant Reformation shifted the balance of churchly power.

This is true of England in the age of efflorescences after the Norman Conquest of 1066. We begin with the Norman Conquest because this act of barbarism established a relatively cohesive political and fiscal architecture that would eventually be challenged but never wholly overturned during baronial wars, peasant revolts, the English Civil War (1642–1651), and the subsequent political settlement commonly referred to as the Glorious Revolution (1688) (Garnett 2009; Thomas 2003a, 2003b, 2007).[8] After the Battle of Hastings (1066) and the subsequent and violent pacification of rebellious populations, William I assumed the throne and through feoffment, distributed lands to the continental warriors who helped him claim allodial title over all of England. Virtually the entire English aristocracy fled or were killed and their lands expropriated by William and his invasive force. The next phase in the extension of William the Conqueror's power was to command a survey of his new kingdom. Completed in 1086, the Domesday Book provided royal power with information on "landholders, their tenants, the amount of land they owned, how many people occupied the land (villagers, smallholders, free men, slaves, etc.), the amounts of woodland, meadow, animals, fish and ploughs on the land (if there were any) and other resources, any buildings present (churches, castles, mills, salthouses, etc.), and the whole purpose of the survey—the value of the land and its assets, before the Norman Conquest, after it, and at the time of Domesday."[9]

Assessing the lay of the land and much of its monetary value in pounds sterling was facilitated by the fact that King Offa of Mercia (r. 757–796) had gradually laid the 'foundations for a money economy' by adopting the pound as the official currency and opening up new mints to satisfy the demands for coinage used for 'war, tribute and trade' (Davies 2002, 127 and 130). By 928 King Athelstan introduced the idea of a uniform currency in the Statue of Greatley, though the principle was only truly enforced after 1066 (Davies 2002, 130). Thus, well before any other continental power, England had a uniform currency and a standard unit of account, and the Domesday Book helped establish a national system of taxation.[10] By the tail end of the eleventh century, the king and his family owned 17 percent of the land, bishops and abbots 26 percent and about 190 Norman invaders, 54 percent of the land as tenants-in-chief.[11] Thus, ownership over the primary source of wealth—photosynthetic energy derived from the land—was highly concentrated. Moreover, the monarch also owned the mints and had the sole power to open up new mints and coin money.

Since money was thought to be the metallic substance of silver (and later gold), this set very strict limits on the money supply— a major problem for the expansion of capitalism, as we will discuss momentarily. We should also note that ownership over the money supply did not mean that the monarch was all powerful. Once royal authorities spent the coinage into the economy, they could receive a limited amount of money back through taxation, but what could be taxed and at what level was, in large measure, subject to custom and the ability of the king's subjects to resist new and unwelcome taxes. Regents could also call back money in the form of fines and through the sale of portions of their royal estates. The overwhelming majority of this spending went to fund war, which was the king's sole responsibility and one of the reasons why successive monarchs were in constant need of money.

Besides the creation of coinage and taxation, the royal family's income consisted of revenues from royal estates and the sale of titles, charters, or patents. Royals also invested in pirates and plunder on the high seas once it was proven a profitable enterprise to loot Spanish vessels returning to Europe with Latin American silver and gold. Diplomatically, these ventures could not be officially sanctioned by the regent and were publically condemned. Privately, however, they were celebrated. For example, Queen Elizabeth (r. 1558–1603) invested in Hawkins's private slave-trading ventures in the Caribbean and his cousin Francis Drake's circumnavigation of the world by plunder. The return on the queen's investment (and those of other investors) was 4,700 percent (Lloyd 1984, 9). So although the regent controlled the mints and juridically owned all of the physical land of England by right of conquest, built into the social reproduction of class rule in this agrarian economy was the possibility for ever greater financial power to fall into the hands of the king's more privileged subjects who received corporate monopolies, patents, or charters for inventions or commercial enterprises at home or abroad.

Beginning in 1066, there is little doubt that the countryside was gradually being monetized by the process of commuting rents into monetary payments and by taxation on landholding. As Dyer notes, '[T]he demands of the lords and the state for cash' forced peasants into the market to earn money by selling their agricultural produce or petty commodities (1997, 42). To this we could add the multiple monetary fines that could be applied by authorities across the population (Dimmock 2014, 101ff). Although these pressures to find money did not make all peasants completely dependent on the market for their social reproduction (many still had access to the land and could produce for their immediate needs), it did make the market for goods and labour and the emerging price system more important to everyday life. Peasants with more substantial holdings also gained money by selling surpluses on the market. Money could be used to buy land; animals; and equipment for farming, cloth, buildings, firewood and food (Dyer 1997, 42). But by the sixteenth century, both fines and rents started to skyrocket in some areas of England.[12] This was largely because surveyors assessed the differential power of the land to yield expected future money and landholders desired to accumulate more money, not simply to pay taxes, but to garner more money and social prestige (Tawney 1912, 119). This added tremendous and increasing pressure on tenant cultivators, and often their waged workers, to find more money by increasing productivity or the desirability of products that would go to the market. On this point it is worth quoting Clay at length:

> Nevertheless rack renting was probably the more disliked because in the sixteenth century it was something of an innovation, at least as far as peasant farmers were concerned, and because it was often introduced where a landlord had succeeded in converting copyhold land into demesne. Sometimes, indeed, the change to rack rents forced the poorer peasants to give up their farms altogether, whether or not this was the landlord's intention, for if their holdings were too small to pay the rent demanded and to feed their families, and no subsidiary sources of income were available to them, they could not carry on. For those who stayed, however, the economic consequences of the two practices were much the same. A man faced with the need to raise a large fine could either save up the necessary sum in advance, or he could borrow it and gradually discharge the debt over the years that followed, but either way he would have as great a need to increase his annual money income as did the man with a rack rent to pay. This he might be able to do partly by devoting more of his own or his family's time to subsidiary employments or to wage labour, but for most *it was also likely to mean producing more for sale*, and since the possibilities of increasing total output would probably be very limited on a small family farm, *this meant marketing a larger proportion of what he produced at the expense of his own consumption. Higher rents and fines thus meant harder work, or a lower standard of living, or both. They also meant that the balance between farming for subsistence and farming for the market within the peasant sector as a whole was bound to tilt in favour of the latter*. (Clay 1984, 90, emphasis added)

In sum, at least by the 1500s, rack renting and increasing fines were leading to rural expropriations either through the failure to generate enough income to pay rents or by going into debt to pay fines, rent, or taxes. Price was already a weapon of the powerful. But these pressures were not the only mechanisms at work leading toward a deeper dependence on the market for social reproduction and survival. Competitive pressures to produce and monetize production were combined with waves of enclosure that effectively abolished customary rights to common fields and led to a depopulation of the rural countryside and widespread pauperization (Neeson 1993). The primary reason for the first waves of enclosure was the logic of differential accumulation as landlords realized that English wool could be sold on the burgeoning world market at considerable profit. Up until this point, not many English products were desired or needed by foreigners, but there was a perceptible demand for English wool on the European continent given the colder climes. The response was to convert more arable and open fields into pasture to graze sheep—a process that lasted from the fourteenth century until the seventeenth, and one that heralded the decline of commoners, the creation of paupers and the concentration of ownership in ever fewer hands (Tate 1967). As Cipolla notes, by the end of the sixteenth century, 'cloth accounted for about 80 percent of the total exports of England' (1977, 38 and see 202ff). As Thomas More depicted in *Utopia* (1516), it was a time when sheep devoured men and the profits of the trade went 'into a few rich men's hands'.

Even though some minor efforts were put forth by earlier Tudor rulers, by 1601, the social calamity of pauperism could no longer be avoided. Elizabeth I introduced the Poor Relief Act with new amending acts following on in the next century and culminating with Speenhamland in 1795 as the problem continued. Indeed, Wilson (1965), using Gregory King's statistics, argued that by the end of the seventeenth century, 1.3 million out of a total population estimate of 5.5 million were considered paupers and cottagers, and another 30,000 were considered thieves, beggars, or vagrants. 'At a conservative estimate, a quarter of the population would be regarded as permanently in a state of poverty and underemployment, if not total unemployment' and during depressions, 'something nearer a half of the population' (Wilson 1965, 231). As Polanyi remarked, these two centuries are known for a bevy of pamphlets that try to explain the origins of the poor and what should be done about their growing numbers. A number of answers were suggested at the time—many absurd—but Polanyi argued that the real reason for the rise in pauperism and poverty was the waves of enclosure (and we may add fines and rack-renting) that dislocated peasant cultivators from the land and left them with no choice but to find wage labour. The problem was that in both the burgeoning towns and cities and the rural countryside, mass unemployment prevailed so that the dispossessed and the descendants of the dispossessed could not find paid work (Polanyi 1957, 90–2).

Compounding the expropriations of the centuries-long enclosure movement was a series of laws that further restricted English subjects from accessing woodlands in order to socially reproduce their families. This culminated with the Black Act of 1723. The act was the ruling-class response to two groups of organized poachers who blackened their faces and invaded the lands of the rich to steal and kill deer. The act enumerated more fifty criminal offences that were punishable by death, from fishing to hunting hares (Thompson 1990, 22–3). As Perelman (2000) argues, these restrictions and many other laws were a class-based project directed at the lower classes of society. He argues that it was well understood by the classical political economists emerging around the seventeenth century, as well as the ruling class they largely wrote for, that if the lower classes had access to land and other natural resources, they would only work as much as was necessary for their subsistence. The rest of the time they would use for leisure. The poverty and hunger that resulted from dislocating the lower orders from the land acted as a disciplinarian or spur to work. As Arthur Young declared in 1771: "[E]veryone but an idiot knows that the lower classes must be kept poor, or they will never be industrious" (Thompson 1966, 358). Thus the power of taxation, the persistence of fines and escalating money rents went hand in hand with waves of enclosure and the destruction of customary rights that forced the lower orders to rely more and more on markets for their social reproduction and survival.[13] The corollary of this process was the ever-greater concentration of land ownership as 'enclosure impoverished twenty small farmers to enrich one' (Neeson 1993, 22). Indeed, as Overton notes, by 1850 '[l]andlords owned 75–80 percent of the farmland of England' (1996b, 204).

But we cannot remain in the rural countryside if we want to understand how the magnitude and universalization of capitalism as a mode of power emerged. Before we leave our engrossing landlords and dislocated commoners and cultivators, at least two things are of note. First, there is virtually no doubt that the changes in social property relations were important for the development of capitalism insofar as they monetized the countryside and commodified and commercialized agriculture. But the dislocations were not primarily about creating a wage working class and a property-owning class, but an owning class and a market-dependent nonowning class with the *potential* to labour for others in return for wages. And although it is undoubtedly true that some did find paid work, the immediate result of rural expropriation was mass pauperization and widespread poverty combined with a mix of poor-relief and punitive and disciplinary laws well documented by Marx in section 8 of *Capital*. Second, political Marxists find the origins of capitalism in the social property relations that emerged in the countryside between improving landlord, capitalist tenant farmers under pressure to meet rents and ensure profit and rural wage labourers working for the capitalist tenants who presumably were producing more agricultural goods in monetary value than what they were paid in wages. There is perhaps little doubt that rising land valuations and the profit motive compelled greater productivity and exploitation of labour, but even if this is historically accurate and nothing else mattered, what you would end up with is an agrarian capitalism based on the capitalization and commodification of agricultural production. Even with the agricultural revolution, which Overton (1996a) convincingly dates from 1750–1850, the capacity for capitalization would have been strictly limited and capitalist social reproduction would have remained limited by renewable sources of energy. Furthermore, although agriculture did contribute to 50 percent of the national product, even in 1688, and land, the most common investment, this sector would hardly become the most capitalized (Cipolla 1977, 83). The most gigantic capitalization was far from agriculture, but in the government debt and the state's power to tax, spend, regulate and make war. In this way, changing agrarian social relations are not sufficient to explain the rise and greater universalization of capitalism as a mode of differential power, though they surely remain a part of the story. So although these remarkable dispossessions set England apart from other nations, we have to look elsewhere in the countryside to find the one energy source that would propel capitalization forward: coal. And from the bowels of mine shafts we have to move to the emerging cities and the heart of state violence and finance: London.

THE AGE OF CARBON ENERGY: A CONFLUENCE OF EVENTS IN ENGLAND

Long before the invention of the oil barrel, there was the coal chaldron.[14] If the oil barrel can be taken as the leading symbol of carbon capitalism, then the coal chaldron should be taken as the sign of its emergence. Coal was used by many civilizations, but something happened in England that made it *dependent* on the increasing extraction of coal. There is evidence of coal being fired for heat and light in Roman England (43–409) and although it may have fallen into disuse for a time due to the superiority of wood for heat and light (wood is less noxious), it was certainly known about in London by 1228. Queen Eleanor is also alleged to have fled Nottingham Castle due to coal fumes, and by 1257 Cipolla tells us that there are clear mentions of coal being shipped into London from the north. Due to its sea travel from the mouth of the Tyne in the north to the docks of the Thames in the south, the combustible rock was called 'sea-coal' (Cipolla 1977, 2010). Although coal was not just found in the north of England, before the canalization and the creation of better roads in England in the eighteenth century, transporting coal to London on land was prohibitively expensive. It was much more cost efficient to travel down the eastern coast and up the Thames. At least by 1640, a pamphlet declared that over 200 sail ships carried coal to London and many more to coastal towns (Turner 1921, 3). There is also evidence of coal trade with continental countries as well.

It is difficult to say at precisely what time coal became absolutely indispensable to the social reproduction of England, but there can be little doubt that by at least the seventeenth century the carbonization of everyday life and industry was in considerable motion. Certainly by the end of the seventeenth century, Turner (1921, 4) argues that coal was indispensable to a number of industrial processes, from iron making to brewing, and Wrigley notes that coal use in England had made the country noticeably different from the political communities of the continent, if not the world, by 1700 (Wrigley 2010, 64 and 111). In fact, just as the US Navy protects sea lanes for oil tanker transport to ensure the social reproduction of the American way of life, so, too, was London's navy highly concerned to guard the coal fleets of Newcastle by at least 1629.[15] After the Navigation Acts passed in 1651, there was considerable worry that the Dutch might retaliate by burning coal ships en route to coastal towns and London.[16] The English won the First Dutch War (1652–1654), however, and fears of cutting off London's energy supply never materialized. The naval war solidified the Navigation Acts against the Dutch and the coal trade amplified as the seventeenth century continued.

As it turns out, however, the major danger that threatened the use of coal energy did not come from outside England, but from inside. Called the hostmen or hostelers because they were originally responsible for entertaining foreign merchants and overseeing their conduct, by 1600 this group managed to have a practical monopoly over the coal trade in Newcastle. They gained control over wayleaves and wharf rights such that no coal could pass down the Tyne and out to sea save through their operation (Turner 1921, 7). In return for additional tax on coal, Queen Elizabeth made their de facto monopoly legal, thereby granting them more power over the trade. It was not long before they expanded their enterprise by purchasing nearby coal mines or by paying dead rents to their owners. In this latter practice, a rent and a royalty would be paid to the owner in return for not independently mining the coal.

Other acts of monopoly and sabotage that presage the oil industry of the late nineteenth century include (1) harassment of competitors and the flooding of their mines, (2) securing more wayleaves so that competitors could not gain access to the Tyne to sell their coal, (3) using their monopoly privileges to force taxes on coal traders in other regions like Sunderland, (4) making exclusive agreements with the masters of ship or acquiring a stake in the ships themselves, (5) ensuring that the workers did not combine to elevate their wages and conditions of work (Turner 1921, 8ff) and (6) in the eighteenth century, imposing fines on individual collieries who exceeded production quotas (Cromar 1977, 86).[17] Coal and the threat of elevated prices during this period seemed to be so important that by 1655 a pamphlet entitled *England's Grievance Discovered in Relation to the Coal-Trade* appeared complaining of the hostmen's tactics:

> Ingrossers of all Coals, and other commodities, into their own hands . . . what they cannot do by force of their Charter amongst themselves, against any private person opposing, then by Combination ruin them at Law, by their Delatory Plea, and out-pursing them, to the high dishonor of God . . . They will not suffer any of the Coal Owners in any of the two Counties to sell their own Coals, but the

> Owners must either sel their Coals to the free Hoast-men, at what price they please, and then all ships must give them their own price, or get none. This it is which makes coals so dear: they either hoard or sell at excessive rates, and so reduce the people to miserable condition. (Gardiner cited in Turner 1921, 8)[18]

In addition to the acts of sabotage noted earlier, what this passage illustrates is that arguably the first energy company in the world was founded not on productivity for productivity's sake, but on the ownership and strategic control of production for the sake of profitability—a key tenet of the capital-as-power approach to political economy. This strategic limitation allowed the coal industry to expand in scale over time (sabotage is almost never 100 percent of production) by ploughing monopoly profits into the development of more mines and more equipment. By the first half of the eighteenth century, we find 1,200 ships in operation with 15,000 men responsible for navigating the coal to paying ports and an estimated 100,000 people employed 'above the ground and under it' (Turner 1921, 6). A complete monopoly capitalist energy complex was emerging on a vast scale, only to be gradually undone in the latter half of the eighteenth century as canals, and later in the nineteenth century, railways opened up additional coal mines to markets across Britain. In sum, early coal 'capitalists strove by various devices, particularly by combination, to destroy competition, monopolize markets, and fix prices as they desired' (1921, 23). As we will see, this example is hardly the exception but the general rule of carbon capitalism.

As important as coal would become to the creation of a British-led world order by the nineteenth century, it is of crucial importance to consider another form of energy trade at the time: the transatlantic slave trade. If we exclude Ireland and the perennial attempts to win back parts of France, England embarked upon colonization much later than its European counterparts (Canny 1998; Lloyd 1984). Monarchs were hesitant since missteps abroad could cause diplomatic trouble with continental powers, not to mention expensive wars that had to be shouldered by the royal treasury. So charters were only granted to companies with a reasonable prospect of achieving a return on investment and whose activities were unlikely to strain the royal finances by provoking the wrong parties. These were all commercial companies capitalized on the basis of an expected future profit and as Blackburn reminds us, by 1614, 'no fewer than a third of all Members of the House of Commons were involved in colonial projects of one sort or another' (2010, 220). In due time, English merchant companies and gentry in search of landed wealth abroad gained control over some of the islands in what became known as the West Indies and parts of the eastern seaboard of today's United States.

Sugar was the most important crop of the West Indies and tobacco the most important crop of the nascent Virginia colony. Both were reasoned to bring in revenue streams through exports to England and the European continent. The problem was that in an age before fossil-fuelled agriculture, both crops were extremely labour intensive to produce on a large scale. The solution to the 'problem' was the capitalization of human energy in the form of the transatlantic slave trade. In 1600, a group of merchant adventurers huddled around Prince Rupert made a deal with King Charles II to found a company to exploit West Africa. During the Interregnum, Rupert had been to the west coast of Africa and heard many stories about gold mines in the interior. The new joint-stock company was to be capitalized on the basis of finding these mines and exploiting their gold. The King was promised two-thirds of any gold found, prompting him to offer five ships from the Royal Navy to accompany the first voyage (Blackburn 2010, 249ff; Zook 1919, 148ff). The success of the company in finding gold and other tradable goods was lacklustre until it was realized that the colonies needed labour and West Africa could be a lucrative source of supply at profit.

In 1663 a new joint-stock, The Company of Royal Adventurers of England Trading into Africa, was subscribed to. The company received more privileges and advantages in its charter, and the firm turned to slavery as a way of making returns on their investments. However, their monopoly on the African trade incensed the Dutch who had largely gained control of the trade during England's Civil War (1642–1651). Wars with the Dutch over commercial supremacy soon followed the incorporation of the company, crippling it with financial difficulties. Thereafter a new monopoly called the Royal African Company was capitalized at £111,100, which looked wholly to the slave trade with the sugar and tobacco colonies of the United States and the West Indies for profit. Leading figures such as John Banks, Josiah Child and John Locke subscribed to the company, as did members of the royal family. From its period of incorporation until the end of its monopoly in 1713, the company bought 125,000 slaves on the African coast, lost a fifth in transport along the 'Middle Passage' and sold 100,000 to planters in the West Indies. Slaves would then be set to work under gruelling conditions producing sugar for export to England/Britain and re-export to Europe (Blackburn 2010, 255). Sugar consumption in Britain would not be widespread and cheaply available until the nineteenth century. But Mintz argues that its production not only contributed to the growing knowledge concerned with worker control, but also was a source of cheap food/fuel for energizing the emerging industrial working class (Mintz 1986).

In capitalizing slavery, the bigger prize was, of course, the Spanish *asiento*, the monopoly right to trade African slaves to the much larger Spanish colonies of the Caribbean and Central and South America. Moreover, due to the control of Latin American mines, Spain was in possession of the world's largest store of silver and gold at the time—metallic substances necessary for international trade—particularly with East Asia. Charles II tried to obtain the *asiento* for the company but to no avail. It would take the defeat of Spain in the War of the Spanish Succession (1701–1714) to wrest the right from the Spanish throne in 1713. The new political regime handed over the *asiento* to the South Sea Company, which would now hold the monopoly on the British slave trade. To satisfy the investors of the Royal African Company, the company was given a contract to provide a certain amount of slaves to the South Sea Company. In sum, if the hostmen can be considered the first energy monopoly in fossil fuels, then the South Sea Company can be conceived of as the first English enterprise dealing in enslaved human energy. As we will discuss momentarily, the South Sea Company was also intertwined with the largest capitalized entity of the period: the British state through its 'national' debt.

The key question is why there was such a demand for coal in England and slaves in English colonies.[19] I will address coal first and then the need for slaves. Although in the past historians have disputed this question, in the current literature there is little dispute that at least by 1600, if not before, there was an energy crisis in England (Brinley 1986; Cipolla 1997; Fouquet and Pearson 1998; Goldstone 2002; Malanima 2006; Nef 1997; Podobnik 2006; Smil 1994, 159; Wrigley 2010). What exactly caused this energy crisis is disputed, but it is fairly certain that a number of factors were involved that all contributed to deforestation and increases in the price of timber. These included population growth, the production of charcoal for industrial purposes, timber for domestic heating and construction, a massive increase in shipbuilding after the Navigation Acts (1651) and Cromwell's expansion of the navy, the rebuilding of London after the Great Fire (1666) and royal wood monopolies. Although it may be difficult to say which factor or factors played the greatest role, the evidence that there was a verifiable crisis is in little dispute, given the historical record and the mounting cost of wood and charcoal. For example, as early as 1548, royal authorities 'ordered an inquiry into timber wastage and deforestation' (Cipolla 1977, 180). By 1670, the Marquis de Seignelay was informing his father, French Minister Colbert, that England did not have enough wood to produce armaments and that they were purchasing cannon from Sweden (Cipolla 1977, 180). Earlier in the same century, the English chronicler Edmund Howes wrote that "there is so great a scarcitie of wood through the whole Kingdom, that not only the Citie of London, all haven-towns [ports] and in very many parts within the land, the inhabitants in generall are constrained to make their fires of sea-coal or pit-coal, even in the chambers of honourable personages, and through necessitie, which is the mother of all arts, they have of very late years devised the making of iron, the making of all sorts of glasse and burning of bricke with sea-coal or pit-coal" (cited in Cipolla 1977, 209).

There is considerably more evidence, but it is clear that England increasingly turned to coal, given a shortage of timber and the rising cost of wood and charcoal as fuel and building materials. But as mentioned earlier, our analysis should not stop at scarcity but should inquire into the reasons why deforestation was so prevalent and why a petro-market civilization emerged founded on coal. It will be my contention that geopolitical conflict and ruling-class warfare played the most crucial role in the shift to coal. But before turning to the role of organized violence and the need for armaments and ships, we must address the question of why there was a twin demand for the energy of slaves.

The transatlantic slave trade was capitalized on the basis of an entire apparatus of violence that would see African men, women and children captured, shackled, jailed, physically assessed, transported, physically and mentally abused and finally driven to work on plantations thousands of miles away from their ancestral homes. But this entire biopolitical apparatus of profit making would hardly have become so gigantic and so long term without an incessant and constant demand for slave labour to work the sugar and tobacco plantations in the Americas and Caribbean. Zook provides some of the main reasons why the English and other European powers resorted to slave labour: "Slaves and their children after them were chattel property for life. The danger of rebellion was very small because often the slaves could not even converse with one another, since they were likely to be from different parts of Africa and therefore to speak a different dialect. Finally, neither the original outlay for slaves nor the cost of feeding and clothing them was great, and therefore salves were regarded as more economical than indentured servants" (1919, 206).

In other words, slave property could make up for a shortage of labour power because it could be more easily controlled than other forms of servitude. But it should always be remembered that we are not talking about some arbitrary shortage of manpower. Colonists could have farmed for subsistence by themselves, in groups cooperatively, or with their families. In this sense, there was no labour shortage. But the intent of the colonist planters was not to farm for subsistence, but to plant and harvest for profit so that they could live like country gentlemen on large estates. Often, foreign plantations were capitalized by absentee owners living in Britain. Many of the same absentee owners also owned slave property on their foreign plantations (Draper 2010; Thorne 2012). In sum, slavery was capitalized by the wealthy and want-to-be-rich for profit, right down to the last crack of the whip and drop of blood.

It is, of course, true that Britain was the first to ban the slave trade on seagoing vessels in 1807 and then abolished slavery in the British Empire in 1833. But as Sherwood (2004) points out, there was a massive difference between the legislation and facts on the ground. In multifarious ways, from trading with slave-owning countries to financing shipping and insuring slave vessels, British capitalists continued to capitalize on the trade in human energy well after the early nineteenth century. Moreover, as Draper (2010) has demonstrated, absentee slave owners in Britain and British slave owners living abroad received incredible compensation for their human 'property'. This compensation, he argues, was essential to the gradual dismantling of slavery in the British Empire. By the Slavery Abolition Act of 1833, the enormous sum of £20 million pounds was set aside to recompense slave owners for their losses. Since the compensation was financed by parliamentary debt, the working-class tax payers of Britain essentially paid wealthy slave owners for the loss of their 'property' *plus* interest on a ballooning national debt that was all owned by the wealthy 1 percent of Britain (Ferguson 2006, 195). Moreover, upon receipt of the reward, many proceeded to buy government securities to earn interest on their compensation, thus making even more money at the public's expense, given Britain's regressive tax system (Ferguson 2006, 194). It will not surprise the reader to find out that between 1820 and 1835, 100 members of Parliament (MPs) were found in the compensation records studied by Draper. Two-thirds of MPs during this time were slave owners and the final third trustees or executors of estates of slave owners (Draper 2007, 90).

But the ever-increasing use of coal and the development of a colonial system founded on slavery, violence and cash crops are not sufficient to explain the magnitude of capitalization and the universalization of capitalism as a mode of power. Slavery and coal were deeply and inseparably interconnected with the Glorious Revolution (1688), geopolitical dynastic war, the capitalization of the state and the rise of what Ingham (2004) calls capitalist credit money. Because these events are all interrelated, there is no obvious starting point, but my own preference is to begin with a brief understanding of the monetary situation by the middle of the seventeenth century. As suggested earlier, in Europe and Asia the dominant forms of money were silver and, more importantly, gold. The search for more silver and gold was one of the biggest reasons for the violence of colonialism because amassing these metals and coining them gave royal authorities greater power to command armies and equipment (e.g., weapons, ships, horses), build monumental architecture (mostly of a religious nature), purchase foreign luxuries and capitalize additional income-generating enterprises like the slave trade or trade with the Indian subcontinent. As Bernstein noted, '[G]old would become the preeminent tool in the management of economic power. Gold's strategic role became so dominant over time that the struggle to obtain adequate sources of it would motivate monarchs and nations to great deeds and tragic treachery' (Bernstein 2004, 87). Thus, there is little doubt that there was a drive to accumulate money among the powerful. The problem for the sovereign, of course, was that access to silver and gold was limited by the mines in one's possession, the state of trade and success at plunder. Compounding the problem for sovereign authorities was the fact that once money was spent into the economy, only limited amounts of it would be received in the form of legitimate taxes. Because it was the sovereign's responsibility to pay for war and wars were one of the most expensive ventures undertaken by monarchs, regents were often beholden to the subjects or foreign creditors for financial loans to support their violent undertakings. Loans could also be forced on subjects, but quite obviously, this was not a popular measure, particularly when the king took it upon himself to absolve his debts.

As Wennerlind (2011, 17) has demonstrated, what appears remarkable about seventeenth-century thinkers in England is their preoccupation with the dearth of money as the key problem to solve to improve society and remove the scourge of poverty and pauperism. It seemed clear to English members of the Hartlib Circle—essentially an international information-sharing network—that much more surplus, improvement and economic activity could be generated if only more money could be found to help circulate and mediate exchange. At first, the Hartlibians resorted to the age-old practice of alchemy. Their early 'mental marriage' to money as metal led some Hartlibians to try to turn other base metals into silver and gold. Eventually, however, the Hartlib Circle ushered in an epistemological revolution by breaking from the metal fetish:

> The Hartlibians maintained that the exchangeability of money was determined by people's trust in money's capacity to serve as a pledge and security in market exchanges. As such, they departed from the neo-Aristotelian tradition of conceiving of coin as mediating commerce because it embodied the same intrinsic value as the commodities for which it was exchanged. Instead, they argued that people were willing to accept money in exchange for their goods because they believed that money would enable them to purchase other goods of the same value at a later date. Money's value was therefore determined more by the future than by the past or the present. (Wennerlind 2011, 83–4)

Eventually, this type of thinking became more commonplace and almost certainly was in circulation by goldsmiths before the Hartlibians arrived at the point. Thus, the money supply could be expanded if those involved in market transactions had faith that a certain medium could be exchanged for goods or services at a future date. From a technical point of view, expanding the money supply could be done in a number of ways and throughout this period many pamphlets circulated with proposals (Davies 2002, 256).

Eventually, Parliament settled on the idea of a Bank of England run by private operators. These operators promised to provide the government with a permanent loan of £1,200,000 in return for incorporation, 8 percent interest and an annual management fee of £4,000 (Davies 2002, 260). The bank, established in 1694, was also permitted to engage in commercial operations, and its notes circulated as England and Europe's first credit currency ostensibly backed by an equivalent amount of silver, but in reality, only backed by a fraction of all money in circulation. As Wennerlind notes, the founder of the Bank of England, William Paterson, thought that 'a reserve of 15–25 percent' of silver would suffice to back the notes issued by the bank (2011, 110). Other than the need for war finance, what facilitated this financial revolution was the Glorious Revolution of 1688. The revolution essentially subordinated royal power to Parliament. It was now the king-in-Parliament who had the power to tax, spend and make war. Thus, garnering war finance was no longer the household responsibility of the regent, but of a Parliament filled with propertied men. Thus, one of the key problems of the seventeenth century—the dearth of money in the face of increasing economic potential—was solved by the capitalization of the state's power to tax. The capital-as-power approach has long noticed how important the capitalization of the state has been for the development of capitalism. Far from viewing a distinct realm for politics and economics, the capital-as-power approach recognizes that at the very foundation of capitalism they are fused: "The modern nation state, from its very beginning, was highly dependent on capitalist finance, whilst capitalization was similarly reliant on state power. Indeed, it was this fusion between them which gave rise to the first form of modern capital—the government bond—whose very essence was the private ownership of the government's power to tax. Since then, the overlap grew deeper and wider, with an increasing proportion of capital values depending on, and in turn dictating the nature of key political institutions and organizations" (Nitzan and Bichler 2002, 13–14).

Indeed, to this day the government bond remains the heart of global capitalism and it remains the basis for the entire financial system (Henwood 1997). Without government securities, capitalists would be lost and the capitalization formula thrown into uncertainty given that the interest rate represents the 'normal rate of return'. This financial setup—a private, for-profit, bank funding government war debt by creating more money on the basis of a far smaller amount of silver—also had immense repercussions at the time and despite some alterations over the centuries, for the present financial order. What it meant at the time is that the primary way more money entered the economy was through an interest-bearing, debt-based currency for war spending. Because interest is never created when money is lent, this means there is always more debt in the economy than there is the ability to repay. This arrangement is an automatic trigger for economic growth because in order to repay the debt and interest, more activity in the form of commodification and sale has to take place; otherwise, the system collapses as loans fail and debt is not repaid or serviced.

But these early capitalists were not simply trading shares in the power of the state to tax its subjects and relay part of that money to their coffers. They were also capitalizing the increasing militarization of state power as the ruling class, with some exceptions: clamoured for overseas expansion; and invested in the burgeoning joint-stock companies proliferating in the years after the Glorious Revolution. As Kindleberger and Aliber note, '[B]y 1695 there were 140 joint stock companies with a total capital of £4.5 million, more than 80 percent had been formed in the previous seven years. By 1717 total capitalization had reached £21 million' with the vast majority of the companies aimed at overseas trade (2005, 47). The fusion of state and private power here should not be mistaken: "The joint-stock companies effectively enlisted the assistance of private capital to extend state power in other ways. Enterprises specializing in colonization and trade contributed to national economic growth and provided governmental administration and military protection in foreign territories. Joint-stock trading companies bolstered national power by stimulating the expansion of the merchant navy and its allied shipbuilding and domestic armament industries" (Baskin and Miranti 1997, 63).

Exactly what role war and the preparation for war played in the development of carbon capitalism can be debated, but that it played a decisive and vital role should hardly be in doubt. The best indication of this connection is war spending and the rising national debt, combined with the ways in which these called forth development in the means of destruction and a dependence on imported timber as an energy source for smelting iron ore. Such military–industrial developments also significantly contributed to private industry through spin-off technologies that resulted from state-sponsored research and development (Podobnik 2006, 21; Trebilcock 1969). To recall, a scarcity of timber crisis that elevated the price of wood and charcoal for increasing numbers (and particularly the poor) combined with geopolitical violence for ruling-class accumulation forced Britons to turn to coal as a source of heat energy. There is little doubt that the production of iron artillery for military demands at home and for export abroad contributed to the timber crisis. For example, by 1575 England was producing over 500 tons of iron artillery, and by the end of the century, about 1,000 tons a year. Its blast furnaces (including those of Wales) 'produced some 5,000 tons of iron per annum' in 1550 and '18,000 tons per annum around 1600' (Cipolla 1977, 205 and 208). Because coking technology was not discovered in England before 1709 by Abraham Darby, there was no substitute for charcoal in smelting iron ore and this meant energy imports in the form of Baltic timber. According to Thomas (1986), at least three developments contributed to this unprecedented demand for iron and timber from the 1640s to the end of the century. All were connected to enhancing warfare and profit, directly or indirectly: (1) the Cromwellian revolution that called for a massive extension of the Royal Navy and therefore ships of all rates; (2) the commercial revolution resulting from slavery, foreign trade, and the cash crop plantations of tobacco, sugar and cotton and (3) an agricultural revolution with a growing demand for iron implements (Thomas 1986, 134ff). Thomas summarizes the impact of these measures:

> [T]he Cromwellian revolution, one of the most traumatic events in British history, had consequences which went far beyond the constitutional sphere. It turned England into an aggressive mercantilist power with naval supremacy, at a time when her economy was being transformed by a commercial and agricultural revolution and the expansion of many coal-based industries. This conjuncture of events entailed massive shifts in the nation's demand for timber, iron, and, charcoal, and a powerful investment accelerator effect. The available statistical sources suggest that the annual consumption of iron in Great Britain per head of the population almost doubled between the 1630s and 1680. (Thomas 1986, 137)

But although the events in the 1600s put Britain on a more intensive war footing, it was developments in the next two centuries that would create a more intensive and extensive form of carbon capitalism centred on an ever-expanding capitalization of the state and constant war or preparation for it.[20] As Brewer notes, '[A]fter 1688 the scope of British military involvement changed radically. Britain was at war more frequently and for longer periods of time, deploying armies and navies of unprecedented size' (1989, 22). It is true that first wool, and later cotton, played an incredibly important role in the development of British capitalism. But no amount of cotton cloth, thread, or woollen sweaters and socks could have commanded the level of capitalization commanded by British state power. The government bond backed by taxation, a system of tariff protection, colonialism and slavery and near-constant warfare became the heart of British finance, not shares in textile companies. Moreover, the technology used in the textile industry, which was mostly powered by flowing water, would have been unlikely to inspire the prime mover of carbon capitalism: the steam engine.[21] As Hicks argues: "[Early cotton machinery] fits better as an appendage to the evolution of the old industry than in the way it is usually presented as the beginning of the new. There is continuity between the eighteenth-century development of Lancashire and the West Riding and the pre-Industrial Revolution world. There might have been no Crompton and Arkwright, and still there could have been an Industrial Revolution" (cited in Cipolla 1977, 211).

Indeed, the commercially viable steam engine was not invented by textile manufacturers, but by the ironmonger Thomas Newcomen (1664–1729) in 1710.[22] Newcomen mainly designed, manufactured and sold iron implements to the mining industry. At the time, one of the major problems was that coal and tin mines would flood, therefore making it near impossible for miners to excavate ore and coals. Existing drainage technologies were woefully inadequate and largely relied on manual labour or horsepower. Though it would have additional applications, Newcomen's engine was designed specifically to pump water from mines (Allen 2011, 375). James Watt would later improve the efficiency of the engine and obtained a patent for his invention in 1769. Due to the length of the patent, further innovations were stunted as Watt and his financier Boulton made monopoly profits (Smil 1994, 161ff). It is only by 1870, when 95 percent of textile mills had adopted the steam engine, that they were integrated into an already emergent petro-market civilization where British social reproduction was fast becoming increasingly tied to market dependence and coal-fired heat and mechanical energy (Humphrey and Stanislaw 1979, 38; Malm 2013). The carbonization of daily life was intensifying.

As I have suggested, what appears to have been decisive is the ruling class's taste for empire, global commerce and war. It was this campaign for global supremacy that called forth the mountains of coal, steel and debt that set accumulation on a more violent, global and high-energy–consumption pathway. Military spending from 1714 to 1800 was just over £1 billion sterling with total net 'public' debt for war totalling about £440 million by the end of the period (Jurado-Sánchez 2009). Because the British ruling class did not have a significant land army and concentrated on naval supremacy, it is worthwhile to consider developments in that area. Total fixed capital invested in the Royal Navy was £2.25 million by 1750, whereas for all the 243 mills working in the West Riding woollen industry, total fixed capital was a mere £402,651 even in 1800. As Brewer notes, the 'fixed capital in one of the largest sectors of the nation's most important industry was therefore a mere 18 per cent of the fixed capital required to launch the British navy' (1989, 27). Thus the construction of the Royal Navy was enormous even by the mid-seventeenth century, and by 1688 it was certainly the largest and most comprehensive industry in England (Thomas 2003, 14). The scale of this enterprise is captured by Brewer:

> Naval dockyards were, by the standards of the day, immense enterprises. They were the largest industrial units in the country, dwarfing their nearest rivals, the breweries and the mines. During the War of Austrian Succession, for example, the Portsmouth dockyard employed a workforce of over 2000. By the 1770s the total labour force in naval dockyards had reached over 8000, with half of these men working at Portsmouth and Plymouth. The navy was thus one of the largest single employers of civilian labour in eighteenth-century England. Naval ships and shipbuilding operated on a scale quite unlike that of civilian industry and commerce. Capital and labour were deployed in a manner that was beyond the resources of the merchant or manufacturer. Only the state could undertake enterprises on such a scale. (1989, 28)

By 1763, the Royal Navy achieved naval supremacy, ensuring that commercial interests and colonial investments were more firmly protected from rivals. Naval power further increased in the nineteenth century when steam power and iron were fused to create modern battleships. From 1864, the wooden sailing battle fleet became obsolete as steam-powered battleships made of iron replaced wooden vessels (Grove 2005, 47; Sondhaus 2001). By 1880, there emerged a modern military–industrial complex based on steam and steel and centred on long-term government contracts to army and navy contractors (Podobnik 2006, 27). Thus, undergirding the capitalization of the national debt and the expansion of the money supply was a strong navy policy that supported the globalization of British social reproduction as resources from around the world were sucked into the metropolis by the British merchant marine. This was an incredibly costly affair, as the capitalization of the 'national' debt makes clear. After the Nine Years War (1688–1697), public debt stood at £16.7 million. As wars continued to be waged in the eighteenth century, 'public' debt increased by a factor of 15 to reach £245 million by 1783—the end of the American Revolutionary War (Brewer 1989, 93–4). Not a single firm, nor all firms combined, would reach this level of capitalization until the late twentieth century. Thus a confluence of events, all based around ruling-class strategies of differential accumulation in the face of resistance and intraclass struggles, combined to forge a nascent petro-market civilization heavily reliant on coal to create more capitalizable income-generating capacity. And as more energy came online, the magnitude of capitalization increased, as did the geographic scope of British social reproduction. By the end of the nineteenth century, this could not be clearer, and it is to this era we now turn.

THE EMPIRE OF COAL, COERCION AND CAPITALIZATION

We know that by the 1500s in England shares in joint-stock companies were traded. Though there was the occasional public auction, these trades were mostly private transactions and irregular. Ownership was highly concentrated, typically among Londoners and members of the royal family and peerage. The small number of companies—only fifteen major ones—by 1689 had a total capitalization of £900,000 and all were involved in colonial settlement, slavery, or the trade of luxury commodities (Michie 2001, 15). Up until 1700, land remained the primary source of investment and wealth, but during the eighteenth century, developments would lead toward a more permanent securities market, eventually culminating in the London Stock Exchange in 1801. As astute readers can already guess, the basis for the London Stock Exchange was not the joint-stock corporation, despite their presence. The London Stock Exchange was essentially founded on the capitalized permanent debt issued by Parliament to finance war and the constant preparation for war. By 1815, the 'national' debt stood at £744 million held by 250,000 investors, mostly from London (Michie 2001, 53). The British ruling class also capitalized the 'national' debt of foreign governments seeking to raise money in what was fast becoming the world's preeminent financial centre.

The chaos of the French Revolution (1789–1799) and the Napoleonic Wars (1803–1815) put an end to the debate on whether London or Paris would rule international finance throughout the nineteenth century. Thus, the state's power to tax, spend and make war continued to be the most capitalized institution, generating differential income streams to service interest payments. There are at least two factors that explain this other than the organized force and the differential power of the state. First, after 1720, Parliament banned the formation of joint-stock companies unless explicitly approved by the Crown and Parliament. There is considerable debate on why the so-called Bubble Act was passed. Although some have suggested that the law resulted from the excessive speculation in the shares of the South Sea Company, others suggest that the effects of the 'irrational exuberance' was hardly widespread across society and the rapid rise and sudden crash of the stock only affected a few wealthy owners (Davies 2002, 265ff; Hoppit 2002; Landes 1998, 257; Michie 2001, 17; Neal 1990, 109). It is likely that the actual reason for the legislation was to protect the South Sea Company and other monopolies from competition, given the South Sea Company's involvement with helping to service the 'national' debt (Kindleberger and Aliber 2005, 47; Wennerlind 2011, 12).

Whatever the precise intention, the law did seem to have the effect of sabotaging competition from additional joint-stock companies while promoting small businesses and family-run partnerships. The law was not repealed until 1825, leading to a flurry of new company proposals, with some successful and others going the way of the dinosaurs. By the 1840s, the capitalization of state power remained central to carbon capitalism, but a craze for railways triggered a gold rush in capitalization. By 1853, total capitalization on the London Stock Exchange was £1.2 billion, three-fourths of which was still government debt (Michie 2001, 64). But now coal-fired railways were fast becoming the second most capitalized sector on the exchange after debt. From 1800 to 1860, Britain produced more coal than the entirety of Europe, and now more of it was being used to move people and goods both at home and in the empire. Only in the second half of the nineteenth century would some nations inside and outside Europe catch up (Podobnik 2006, 29). The rail and cart originally emerged as a means of transporting coal out of deep pits (Smil 1994, 160). When it was realized that a steam engine could be used to mechanize the process, capitalists promoted steam engines, requiring more energy and iron and furthering the carbonization of social reproduction in Britain. By 1850, the national railway system was completed and so, too, were a 'series of alliances and amalgamations which created a small number of highly capitalized railway companies whose securities offered a secure and predictable rate of return' (Michie 2001, 116). The steam locomotive was also integral to the British and European colonial project and a form of more globalized social reproduction as products from around the world were brought to their imperial centre:

> The railroad as such had enormous territorial implications; it normally required the backing of the host government, and a territorial concession with financial guarantees was usually needed to exclude competing lines, ensure profit and attract the necessary long-term investment. In Asia and Africa, moreover, imperial intervention was often called for to stabilize political conditions for successful railway operations. Steel rails had a capacity for transforming the societies through which they ran and for spreading imperial influence in their domestic affairs, which often provoked anti-imperial reactions and involved European interests in local crises. The locomotive clearly had a unique propensity for integrating and annexing territory, for monopolizing its resources, and for preempting the future of great stretches of country. All these implications, it is suggested, gave rise to a distinctive type of railway imperialism, which added a new dimension to European expansion and projected it to a higher pitch of intensity over a vastly extended range. (Davis et al. 1991, 3)

The process of company formation and amalgamation was facilitated by The Companies Act of 1856, which 'provided an even more conducive atmosphere for marketing of corporate securities by limiting liability in the event of bankruptcy for those who purchased equity shares of incorporated businesses' (Baskin and Miranti 1997, 130). From here, the ownership of differential power over the social process would became increasingly commodified, with 500 stocks available for purchase by 1853 and over 5,000 quoted on the London Stock Exchange on the eve of World War I (Michie 2001, 95). At the same time, like today, ownership of these new income-generating entities was highly confined to the few. As Michie notes, only 0.8 percent of the British population could be said to be classified as investors, whereas this number only increased to 2.2 percent by 1913 (Michie 2001, 72). This is just another way of saying that a small coterie of families and individuals owned the dominant capital of the era, profiting from the carbonization of social reproduction; the discipline and control of a waged working class who did not get the right to unionize until 1871 and, for males, full voting power until 1884; the constant threat of unemployment; and what was fast becoming a renewed imperial quest in Africa increasingly informed by a newly emergent racism based on pseudoscience. But the geographic extent of Britain's differential ownership and social reproduction extended beyond Africa:

> The plains of North America and Russia are our cornfields; Chicago and Odessa our granaries; Canada and the Baltic are our timber forests; Australasia contains our sheep farms, and in Argentina and on the western prairies of North America are our herds of oxen; Peru sends her silver, and the gold of South Africa and Australia flows to London; the Hindus and the Chinese grow tea for us, and our coffee, sugar and spice plantations are all in the Indies. Spain and France are our vineyards and the Mediterranean our fruit garden, and our cotton grounds, which for long have occupied the Southern United States are now being extended everywhere in the warm regions of the earth. (Jevons 1866, 411)

Although Jevons's statement is polemical, it does point to a keen recognition of globalized social reproduction—one that had few benefits for a considerable portion of Britain's working class during the Victorian period, as the work of Mayhew, Booth, Rowntree and Marx and Engels suggest. Wrigley argues that it was really 'only in the second half of the twentieth century' when 'improvement in health, education, and general welfare' were 'widespread, substantial, and sustained' (2010, 209). But by this time, the ruling class had started to construct a more extensive petro-market civilization dependent on foreign oil.

TRANSITIONING TO OIL

As we have discussed, the British ruling class made coal necessary for geopolitical accumulation because these practices—taking away property, land, or goods from others; protecting one's capitalized trade from those who might do the same; and enslaving others—meant having the ability to employ a considerable degree of violence against resistance and opposition. In this sense, there was constant effective demand for war materiel and technological development to facilitate the application of violence or its threat. This ongoing demand, as we have seen, accelerated after the Glorious Revolution when propertied parliamentarians effectively controlled government policy. In the twentieth century, the decision to transition to oil and forms of high-energy social reproduction was also made by the ruling class of Britain in the form of the Admiralty. The main reason, once again, was to ensure British naval supremacy and ruling-class geopolitical accumulation for owners of British and foreign assets. Because the working class are, by definition, nonowners of society's income-generating assets, we can be sure that the major benefits of this belligerent policy of supremacy chiefly benefited, albeit in different degrees and with considerable political contestation, the British 1 percent.

As suggested earlier, the need for war materiel for differential intraclass geopolitical accumulation not only called forth the greater extraction of natural resources, but also considerable scientific and technological developments. In other words, there is no way to disentangle carbon capitalism and the capitalist mode of power more generally from war and the preparation for war.

The British navy was at the forefront of these developments and kept a constant eye out for innovations. One such innovation was the use of what Russians called *mazut* in the 1870s. *Mazut* was the dark thick residue left from refining oil, and the Russian fleet used it to propel their steamers on the Caspian Sea (Engdahl 2004, 19; Mohr 1926, 26). When this development was noticed, the British Admiralty were in the process of converting the naval fleet to run on coal-fired steam engines. But by 1882, Captain (later Lord Admiral) Fisher conducted research and found petroleum-burning diesel engines far superior to coal power. The advantages were unambiguous. Oil propulsion made it easier for the navy to refuel, it required less manpower, it did not produce very visible carbon emissions (unlike discharge from coal, which could be seen by the enemy about 10 miles away), and most importantly, it allowed for quicker acceleration and the consistent maintenance of speed. This latter factor was of increasing importance because of the threat of submarine torpedoes (Engdahl 2004, 19; Goldrick 2014; Kent 1976, 4; Podobnik 2006, 66). By 1909, the conversion of the British navy to oil-fired propulsion was underway, with military engineers overcoming technological obstacles to burning oil in small and large engines (Podobnik 2006, 67). This conversion, with other nations quickly following suit, provided a giant boost to a relatively nascent industry that still had difficulty competing against the prevalence of cheaper coal-based forms of social reproduction. This transition also came at a time when electric light was being introduced, reducing and threatening to eliminate the need for kerosene-fuelled lamps—the commodity Standard Oil and other oil companies had built their fortunes upon (Yergin 1991).[23] But the chief problem for the Admiralty was that Britain had no known domestic sources of oil at the time. Just as the British ruling class had become energy reliant on Baltic timber in the seventeenth and eighteenth centuries, in the twentieth century, it would once again become dependent on a foreign energy source until North Sea oil was discovered and developed in 1969. This transition to oil would have tremendous consequences for the future of war, geopolitical accumulation, and the shaping and reshaping of world order. But it would also have remarkable consequences for capitalization and energy-intensive forms of social reproduction in Britain, our chief concern here. Figure 3.0 charts the capitalization on the London Stock Exchange in comparison with total energy consumption for England and Wales.

Because data are not available from the London Stock Exchange prior to 1963, earlier capitalization figures are taken from Michie's (2001) study of the exchange. The chart is relatively straightforward and shows two giant spikes in energy consumption, the first beginning in 1840 and the second after World War II. By 1975, energy consumption continued to increase, but appeared to reach a plateau somewhere above 8,000 petajoules yearly. From this chart, it would appear that capitalization was relatively negligible until at least 1986, when Thatcher launched her 'big bang' financial reforms. In 1986, capitalization reached £100 billion. But this can be deceiving because capitalization

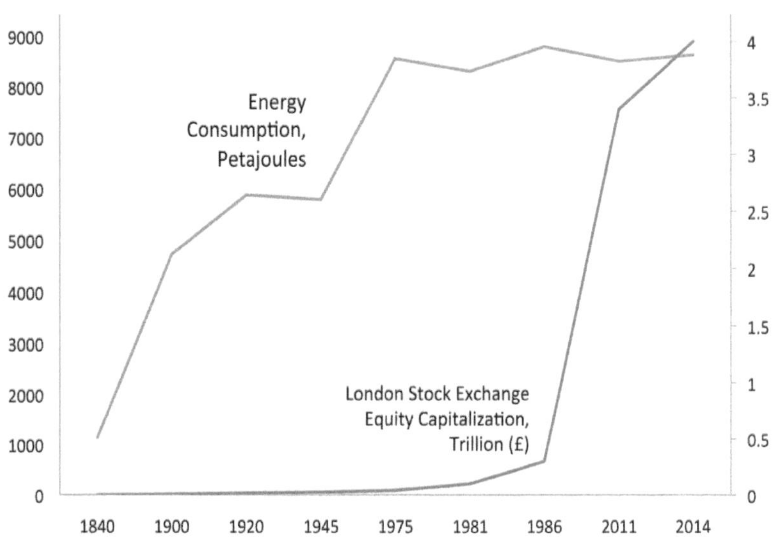

Figure 3.0. UK, Capitalization and Energy Consumption. Source: LSE, Michie (2001), Total Consumption Trends England Wales 1800–2006

was growing steadily throughout the period of growing energy consumption. Still, at least in this stylized chart, a turning point seemed to occur in the 1980s when capitalization took off rather radically. By 1996, equity capitalization on the London Stock Exchange was a whopping £1 trillion, climbing to £2 trillion over the next decade and topping out in 2014 at £4 trillion.

Figure 3.1 takes a shorter view and considers capitalization and energy consumption from 1963, when yearly total equity market capitalization is available from the London Stock Exchange. Once again we find that energy consumption was safely within a band of 8,000 to 9,500 petajoules with no sudden giant spikes throughout the period. Capitalization, as in Figure 3.0, began a meteoric rise in the 1980s, particularly after the 1986 'big bang' financial reforms. There was a major dip in capitalization, which can largely be attributed to the dot.com crash of 2000–2001 when profit expectations did not coincide with actual returns. By 2003, capitalization began to rebound as the War on Iraq began and central bank interest rates remained low. What this chart suggests is that energy is certainly not the only factor that can explain rising capitalization: politics and power matter just as much as increasing the capacity to do work on the natural world. Nevertheless, the chart also suggests that a relatively sustained high-energy consumption above 8,000 petajoules a year is required for increasing capitalization.

Figure 3.2 uses long-term energy consumption statistics from England and Wales and plots this alongside the yearly data series available for the equities on the London Stock Exchange. From this longer-term view, the increasing

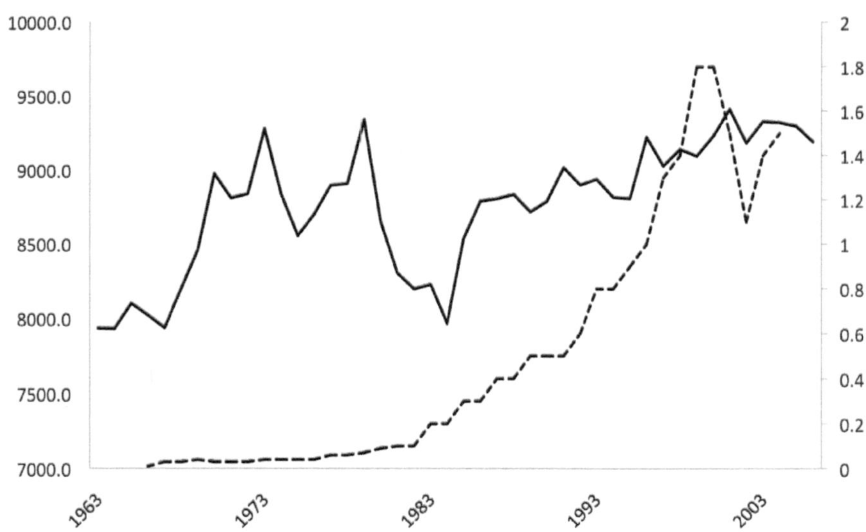

Figure 3.1. LSE Capitalization and Energy Consumption

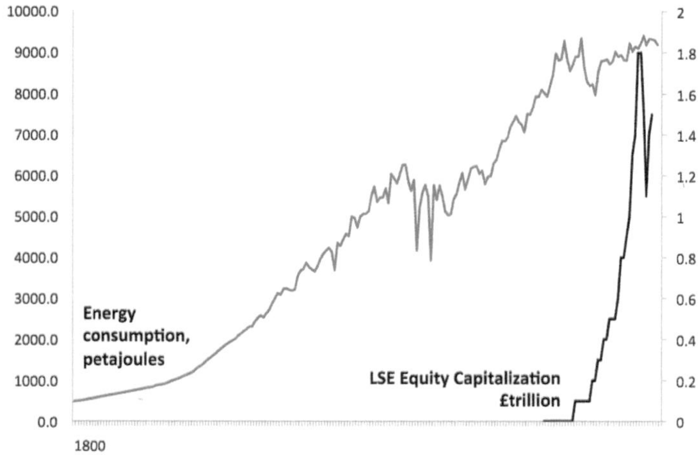

Figure 3.2. Energy Consumption, 1800–2007 & LSE Equity Capitalization, 1963–2004, England and Wales

magnitude of capitalization appears to be ramped up as more and more energy is consumed. Thus, we can make the plausible assumption that if energy drops below a certain level, perhaps 5,000 petajoules a year, we will likely see wild declines in capitalization because there will be less capacity to do work in the global economy.

Last, and intimated earlier, during this period of increasing energy consumption, the United Kingdom's reliance on the world market for social reproduction has intensified. There is no index that accounts for how 'globalized' social reproduction has become or how lives and lifestyles have come to rely on global markets. With this limitation noted, a decent proxy is to consider statistics on imports and exports because an increase in both would indicate greater engagement with the world market and interdependence with foreign workers and supply chains.

Figure 3.3 plots imports and exports from the United Kingdom in the period from 1900 to 1960, albeit with some statistical limitations due to the lack of information mostly during the World Wars I and II. Regardless of the missing data, the trend is clearly upward. The same is true of the data in figure 3.4, which plots UK imports and exports each decade from 1964 to the last quarter of 2014.

Despite some fluctuations, both trends are clearly up, with exports increasing by 2,700 percent from 1964 and imports increasing by 2,488 percent over the period. Clearly this is further demonstration that the United Kingdom is heavily dependent on the world market and energy-intensive production and transportation for its social reproduction. This situation is historically unique

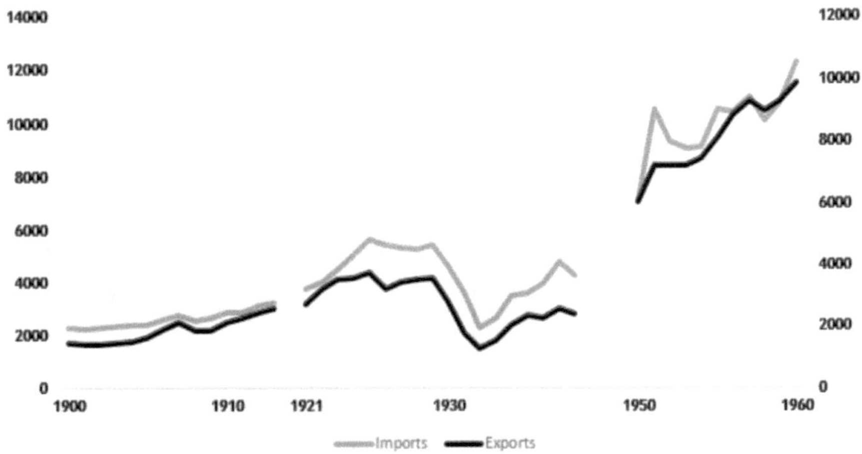

Figure 3.3. UK Imports and Exports, 1990–1960, Total Value in US$ millions. Source: UN Trade Statistics

and in the long run unsustainable. But the United Kingdom is far from alone. Although the construction of a petro-market civilization founded on coal-backed sterling, slavery and the fusion of state and private power began in an internationally interconnected England, its scale would be far outdone by one of its colonial offshoots: the United States. Three thousand nautical miles from England, another confluence of events would take place that would lead to the world's most energy-intensive pattern of social reproduction on the

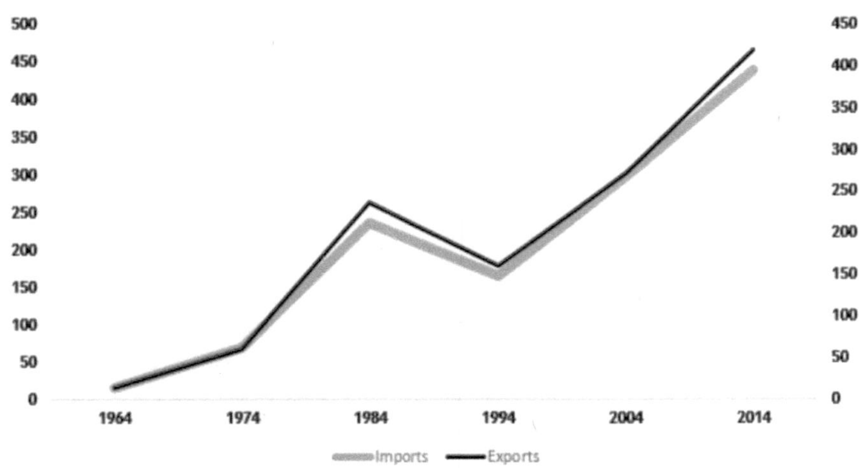

Figure 3.4. UK Total Imports and Exports, 1964–2014, Last Quarter by Decade, £ billions. Source: Office for National Statistics

planet and, like Britain before it, a national security apparatus on a scale quite unlike anything the world had ever seen.

NOTES

1. Marx writes that '[r]aw material may either form the principal substance of a product, or it may enter into its formation only as an accessory. An accessory may be consumed by the instruments of labour, as coal under a boiler, oil by a wheel, hay by draft-horses, or it may be mixed with the raw material in order to produce some modification thereof' (1887, 127).

2. Although I cannot address the entirety of the 'rise of the West' literature here and instead focus on political economy rather than big history, I do address some of the literature in this book. A more critical and lengthy examination, however, cannot be countenanced here, and I leave this work for a future study with more room to elaborate on my critique in greater historical and theoretical detail.

3. Despite the insights of Anievas and Nisancioglu, they remain biblical Marxists wedded to 'mode of production' and 'labour theory of value' theorization with all that entails for historicizing and theorizing capitalism.

4. Data are from the *Financial Times* Global 500, a list of the world's largest firms by market capitalization. The sector breakdown can be found at http://www.ft.com/intl/cms/s/0/988051be-fdee-11e3-bd0e-00144feab7de.html.

5. Interest is considered the normal rate of return because state securities are generally taken to be 'risk free' and therefore the interest rate paid on government bonds can be considered the lowest, or 'normal', return on investment in the marketplace. Of course, although capitalists do hold government debt, they prefer to earn much higher than the 'normal' rate of return.

6. Levy refers to a 1607 pamphlet by John Cowell entitled *The Interpreter*. In the text, the author argues that the Crown's subjects could not truly own property.

7. McKinsey notes that global equities were worth US$11 trillion in 1990 and public debts stood at US$9 trillion. Both figures have exploded since the 1990s, as documented in chapter 2.

8. Davies writes: "The Conquest did usher in a far more complete, a far more standardized system of feudalism, especially in a legalistic, political and administrative sense, than had previously existed in England and it did so at a considerably faster pace than would probably have taken place had Harold, rather than William, been the victor on 14 October" (2002, 135).

9. http://www.domesdaybook.co.uk/faqs.html#4.

10. One pound of sterling made 240 silver pennies. The accounting system, which was adopted from Charlemagne's Frankish Empire, was as follows: 4 farthings = 1 penny, 12 pence = 1 shilling, 20 shillings = 1 £.

11. http://www.domesdaybook.co.uk/life.html.

12. This was called rack renting, or essentially the dramatic increase in the rent demanded by landlords for working their land.

13. These technologies (debt as taxation and debt as fines) would also force colonial subjects to enter labour and cash markets for British colonists in the nineteenth century.

14. Chaldron is the old English spelling of cauldron and was a dry volume measure for coal. Before its weight was fixed by authorities in the seventeenth century, taxation was per chaldron, so coal owners had an interest in packing more weight per chaldron to avoid paying greater taxes.

15. I share Clark's argument that oil (and in the past coal) was not just another commodity like diamonds or fish. As the war veteran Stan Goff has said, 'Oil is not a normal commodity. No other commodity has five US navy battle groups patrolling the sea lanes to secure it' (cited in Clark 2005, 33).

16. The Navigation Acts, passed during the interregnum, were essentially protectionist acts that sabotaged others from participating in trade with the English colonies. The acts made it so that no cargo could be transported in foreign ships from English colonies to England. Moreover, three-quarters of the sailors had to be English or Welsh, and goods from the colonies had to be brought to England first before being potentially resold in European markets (Sawers 1992).

17. Astute readers will notice how these quota restrictions presage the Organization for Petroleum Exporting Countries (OPEC) cartel formed in 1960s to control the flow of oil and, to some extent, its price.

18. A dilatory plea (modern spelling) essentially delays the court from taking action in a plaintiff's case.

19. A separate question we could ask, but that is not explored here due to limited space, is why other political communities did not use coal on the same scale as England until much later.

20. Of course, other European powers looked to bolster their own armed forces. However, Britain had many advantages: the technology to smelt iron ore with coke (a coal derivative), an insular island geography with multiple access points to the sea, and by 1694, a system of credit anchored to the national debt that would allow for an expansion in the money supply beyond silver and gold. Moreover, as Brewer notes, most continental powers focused on building a strong standing army. Britain's priorities were different. They focused on building a strong naval force, thus calling forth incredible resources and massive manpower, and subsidizing foreign troops rather than building up a comparatively sizeable standing army (1989, 26)

21. The textile industry did eventually adopt steam technology after 1830, when capitalists realized that they no longer had to be tethered to flowing water. In this way, the steam engine allowed them to access a pool of labour in the cities (Malm 2013).

22. Newcomen combined the ideas of Thomas Savery and Denis Papin.

23. It is difficult to say in any definitive way whether the oil industry would have survived without the stimulus and technological developments of modern warfare, but there are two things we can be sure of: (1) modern warfare without oil is unthinkable; and (2) the level of destructiveness unleashed on the poor and working classes of the world after the transition to oil has been unparalleled.

4

The Expansion of Petro-Market Civilization in the United States

If William Stanley Jevons was worried about the exhaustion of coal in Britain, by at least the 1920s, leading voices in the United States were fretting about the potential depletion of its domestic oil supply. For example, a leading petroleum engineer of the time could say with confidence that 'no one disputes the fact that our production must be supplemented to a larger and yet larger degree in future years from foreign supplies. America's petroleum problem has therefore taken on an international aspect' (Lewis 1921, 367). Already by the early twentieth century, the United States' 'petroleum problem' was understood in an international context as petroleum deposits were being uncovered and developed outside the United States: "Before 1900, forests of derricks stood above the oilfields around Ploesti in Romania, in Baku on the Caspian Sea, in California, in Texas (after 1887) and in Sumatra (after 1893). Before World War I, oil production had also started in Mexico (1901), Iran (1908), Trinidad (1913) and Venezuela (1914). This rapid expansion led to overproduction and declining prices" (Smil 1994, 168).

At this point in history, Washington DC's concern for oil did not hinge so much on the importance of oil to consumer demand—that world had not been forged yet. The full range of products that could be fashioned with petroleum were hardly known at the time, and it was much later before the automobile fitted with an internal combustion engine became more ubiquitous and available to an emergent middle class in the United States. Until Ford, automobiles were largely the playthings of the rich. Rather, the initial worry was primarily connected to the so-called Great War (World War I) and the ability to wage future international wars. World War I had shown to all who cared to look that oil was essential to 'modern warfare and industrial life' (Lewis 1921, 357; Yergin 1991). The dominant owners of the United States would excel at profiting from both, making the country the world's most formidable petro-market civilization in history.

Although there is some debate on whether China has finally surpassed the United States as the world's largest consumer of energy, the United States remains a voracious consumer of fossil fuels to socially reproduce its energy-intensive consumer society and the largest apparatus of violence in the world. It is the largest consumer of oil at 18.5 million barrels of oil a day, with its closest competitor, China, consuming an estimated 7 to 10 million barrels a day.[1] The United States also far outspends the rest of the world on its military. Of the US$27 trillion spent on the military from 1988 to 2013 by the countries studied by the Stockholm Institute for Peace Research (SIPRI), the United States accounted for a full US$11 trillion, or 41 percent of *all* global military spending.[2] Perhaps not surprisingly, by far the largest consumer of oil in the United States is its martial forces: "[T]he U.S. Department of Defense was the single largest consumer of fuel in the United States, using about 90 million barrels of oil, at a cost of nearly $15 billion. The fuel requirements of the U.S. armed forces accounted for approximately 1.3 percent of all U.S. petroleum demand and more than 80 percent of the federal government's total fuel consumption" (Burke 2014, 33; see also Clark et al. 2010). There have been some attempts to wean the Pentagon off petroleum by finding alternatives, but for the moment, the Pentagon not only defends the global oil trade but as a branch of government, also consumes the lion's share of it. If we consider the US capital market, then it is easy to see that although carbon capitalism is near ubiquitous anywhere fossil fuel consumption is high, the United States is in a league of its own. Not only does the United States have the largest (and continually mounting) public debt in the entire world, its equity markets dwarf those of other countries.[3] The companies listed on the New York Stock Exchange and NASDAQ are capitalized at US$26.5 trillion as of November 2014. The next three largest combined—the Japan Exchange Group, the London Stock Exchange Group and the Euronext—do not even account for half that amount at roughly US$11 trillion. As de Tocqueville noted as early as 1840, clearly there is something exceptional about the United States. But rather than find this exceptionalism in republicanism, democracy, or individualism, this chapter argues that whatever qualitative exceptionalism may be said to exist in the United States, it is an exceptionalism founded on abundant energy sources: slaves and wage labourers, water and nuclear power, wood and fossil fuels.

Yet, it was the discovery and extraction of petroleum that played the crucial role in creating a more intensive petro-market civilization and the US-led world order after World War II. This world order would be built on centuries of slavery, water power, wage labour and coal. However, whereas the earlier discovery and use of fossil fuels helped set the United States apart from the rest of the world, we cannot separate the exploitation of these resources from history, social reproduction and the relations of power that shaped and re-shaped the political economy of the United States. As in the last chapter, my concern is not to trace the precise origin of capitalism in the United States, but to try to uncover how capitalization became inseparable from nonrenewable fossil fuel energy and how the magnitude and universalization of capitalization expanded to levels never seen before. I begin with the revolutionary struggle in the age of efflorescences that created an independent United States. Although there is debate on the reasons for this struggle and some of the historiography has been colored by decades of US propaganda and mythology on the 'brilliance' of the founding fathers, I will make the argument that what compelled the colonists to take up arms against Great Britain was legislation that restricted their ability to accumulate more money and property derived from the land, as well as the chance to form their own legislation that would further differential accumulation. Differential accumulation was not discovered in the United States; it was imported with the first settlers.[4] In other words, by 1775, if not before, British rule was understood to be jeopardizing a business-oriented propertied class interested in aggrandizing their differential power over slaves, native Indians and workers by expropriating and commodifying the land and its products. This would all be underpinned by the federal government's power to tax, spend, dispense land, make war and eventually standardize the currency. That propertied men were involved in creating a republic should not be taken to mean that the ruling class all had the same particular business interests. But although they may have capitalized different enterprises, from the slave trade to textiles, there can be little doubt that those who understood property as an income-generating asset were all interested in one thing: accumulating money. After setting this stage, I then consider slavery, capitalism and the Civil War (1861–1865). Here, I argue that from the capital-as-power perspective, no convincing analysis of the rise of carbon capitalism can be separated from slavery and how the violence and particularly the debt of the Civil War propelled carbon capitalism forward. Just as in Britain, war, finance, slavery, coal and private profit formed a tight and historically intricate relationship. The third section considers the United States in the age of carbon energy and the further industrialization and corporate colonization of income-generating assets by tracing the energy–capitalization–social reproduction nexus. The final section investigates carbon capitalism in the United States after World War II and the constitution and reconstitution of energy-intensive social reproduction and world order.

REVOLUTION IN THE AGE OF EFFLORESCENCES: LIMITS TO CAPITALIZATION

It is too easily forgotten that the United States was founded by joint-stock companies in search of land and the profit that could be derived therefrom. Companies like the Virginia Company and the Plymouth Company were capitalized by investors who expected to profit from land or natural resources in the 'New World'. As Weaver's (2003) research has demonstrated, by the time of colonization, the English elite primarily understood land and natural resources through the eyes of ownership and profitability (see also Wood 2002). Land could be for livelihood, but for the powerful and politically connected, it was a commodity to be improved (largely by slave labour) or sold for money and profit (and sometimes both). Land was to be cultivated and improved, and natural resources such as fish, timber and salt were to be exploited for profit. In other words, right from the start, the colonies were enmeshed in a transatlantic economy of money and ideas about how to accumulate more of it.

The main basis of this economy was a monetary order forged long ago and premised on gold and silver as dominant forms of international money. To obtain gold and silver, colonists without a significant store of specie were compelled not just to provide for their own social reproduction, but also to produce for the world market. Thus converting the land into property and cultivating it with desirable international products allowed them to earn money in order to pay for mortgages, taxes and foreign-supplied goods. There were two chief difficulties to this process of what we could call 'imperial farming'. First, although the natives on the east coast of the present-day United States did practice agriculture (along with hunting and gathering), unlike Europe and particularly England, the 'new world' was still heavily forested and thus unsuitable for *widespread* planting and pasture until trees were cleared. But the benefit of a more forested area meant more timber for iron smelting; heat energy and materials for the construction of forts, ships and homes. As Nye notes, in 'the first 200 years after the colonists arrived, North America lost more woodland than Europe had lost in 1000 years' (1999, 20). Timber would remain a key energy resource well into the age of carbon, when it was eventually replaced by coal as a major source of fuel. The second difficulty that also ended up having an advantageous side to it was the question of whether the native Indians actually owned the land. As Banner (2005) demonstrates by consulting the historical record of this period, many arguments were put forward regarding the theoretical, legal and practical relationship of the natives to the land. Could the English just take the land by right of conquest? Did the Indians really own 'private property'? Could land grabbing be justified on the basis that the Indians were not improving the land in the same way as English cultivators? If land could be sold, who had the right to alienate native land by sale? All these questions and more were asked during the initial colonial encounters. Despite these questions and whatever the theoretical response was at home or in London, according to Banner (2005), the practice on the ground was to treat the Indians as though they did have property that could be alienated. But how, then, did it come to pass that from 1600 to the early 1900s, virtually all of the land in the United States was transferred from the indigenous populations to non-Indians? Banner's work suggests that most of the land was given to the colonists as gifts early on when there were few settlers, purchased from local leaders, or gained in wars that were more often than not provoked by colonists encroaching upon Indian territory and their livelihoods.

But other tactics of cunning, violence and accidental and intended germ warfare were involved as well. First, many of the Indians along the coast died of the microorganisms that the English and other Europeans brought with them from home. On many occasions, new settlers or chartered land companies simply took over the arable land when there were no natives left to farm it (Banner 2005, 29). In other words, soil energy was stolen away by disease. There is also considerable evidence to suggest that English elites in control of colonial policy willingly infected certain native tribes by gifting them blankets contaminated with the smallpox virus (Fenn 2000; Flavin 2002; Knollenberg 1954). Second, another way of decimating the native population in pursuit of profitable land was to enmesh them in webs of debt (Banner 2005, 55). The governments of the colonies, companies and merchants would extend credit to Indians and force them to repay in colonial money or desirable commercial goods like beaver pelts. If the Indians were not able to pay, particularly at a time when the beaver was being hunted to virtual extinction, then land was taken to discharge their debts. Debt was the handmaiden of many dispossessions, and it was well understood as a technology of power (Di Muzio and Robbins 2015). This forced the natives further west into the interior and radically disrupted their less energy-intensive regime of social reproduction that was more or less focused on livelihood, environmental stewardship and community rather than on the accumulation of money and power. Of course, there is little doubt that native tribes and bands often sided with one colonial force or the other to protect their interests. Nor is there historical doubt about the very real fact that some chiefs sought their own private advantage by alienating tribal land by sale without the consent of their tribal members (Banner 2005, 68–9). Third, during wars soldiers received payment and often bonuses in the form of land scrip that could be used to purchase land. Often desperate for money and the means of livelihood, soldiers sold their scrip at steep discounts to rich land speculators who amassed these warrants and later purchased large tracts of land (Ferguson 1954; Mann 2003, 176; Wright 2008, 124). One of the most renowned revolutionaries and first president of the United States, George Washington, was heavily involved in these practices in the Ohio Valley. Indeed, Washington became wealthy off speculating in land for profit (much of it expropriated from native tribes), murdering natives to clear land and employing slaves on his own properties (Friedenberg 1992; Mann 2005; Weaver 2003, 104). Washington belonged to a class of planters who understood that differential wealth came from two main sources: obtaining cheap land and selling it to settlers at a profit or growing cash crops with slave labour for the world market. It was this class that would lead the American revolutionary struggle against the Crown and Parliament in London and set the parameters for continental and then global empire.

As stated earlier, the historiography of the American Revolution and its subsequent constitutional settlement has been much debated. Although we have no space to rehearse the intricacies of the debates here, we are not at a great loss, given our focus on energy, social reproduction and capitalization. Although the American Revolution cannot be explained by one factor and one factor alone, we can think of the Revolutionary War (1775–1783) as primarily the result of leading colonists being dissatisfied with colonial legislation made in London. This legislation either blocked opportunities for the further capitalization of land or frustrated opportunities for profit making in other ventures such as merchant trade (Curtis 2014; Ferguson 1979; Holton 1999).

Although the mounting debts of colonial planters owed to English merchants for conveniences, luxuries and slaves were certainly a grievance before the revolutionary period, it was really only after the French and Indian War (1754–1763) when more general grievances among the wider colonial population started to mount. The war (which was also fought in other theatres than North America) escalated the 'national' debt of Britain and in response, Parliament imposed a series of legislation intended to either raise revenue (The Stamp Act of 1765) or stop colonial administrations and land companies from antagonizing natives by settling west of a line drawn along the Appalachians (Royal Proclamation Act of 1763). Settling could cause more tension and conflict with the natives, and therefore more debt for Britain. The Stamp Act, which was a comparatively mild and progressive tax, so infuriated the colonists that mobs entered the streets and tax collectors were routinely terrorized (Einhorn 2006, 18–9). The act was eventually repealed in 1666, but London maintained that it had the sovereign right to introduce legislation affecting the colonies. Sovereignty meant singular power. The colonists, however, were getting further and further from this position, and its leadership largely held that any legislation without consent was a form of tyranny (Reid 1988, 89). Indeed, despite the apparent contradiction, during the revolutionary fomentation, the colonists juxtaposed the language of freedom and slavery to depict their relationship with imperial Britain. A corrupt Parliament allied to a corrupt Crown threatened to 'enslave' their colonial brethren. The infamous eighteenth-century writer Dr. Johnson summed up the contradiction in the prevailing white discourse when he posed the following question: "[H]ow is it that we hear the loudest yelps for liberty among the drivers of Negroes?"[5]

But liberty did not simply mean that taxation had to be consented to by the colonists. It also meant the freedom to accumulate without limit and this largely meant having access to more land west of the Appalachians for real estate speculation and fertile soil. The Royal Proclamation of 1763 forbade further western settlement, and this enraged both chartered land companies and individual settlers. What made matters worse for the companies was that individual settlers often disobeyed the decree and settled west of the Appalachians. Because the legal apparatus often recognized property in land that had been improved, the companies reasoned that the decree damaged their expected future profits. These companies essentially obtained cheaply priced land from colonial legislatures, then surveyed and carved out acres of smaller individual plots that could be sold to incoming settlers at a profit. Lacking specie, the new settlers were typically mortgaged to the companies and laboured to repay their debts by at least producing some goods for the market (Weaver 2003, 107). This, however, did not stop individuals and families from squatting on western land to gain their own independent, debt-free homesteads. Real estate speculators and companies feared that the claim of pre-occupancy would cheat them out of valuable land when the ban on western settlement was lifted (Breen 1985, 35–6; Holton 1999, 3–38). But although there appeared to be some suspicion that the ban would ultimately be lifted, by 1774, one year before the outbreak of war, the frontier was looking more and more like it would be permanently closed to settlement and land speculation. A series of land reforms and the Quebec Act of 1774 effectively denied western land to men like George Mason (Ohio Company), Richard Henry Lee and George Washington (Mississippi Company), Thomas Jefferson (partner in three land firms) and Patrick Henry (involved in five real estate ventures). Blocking access to new energy sources in land that could be monetized on the US and world market was of fundamental concern to wealthy Virginians who led the revolution (Holton 1999, 35–6; Bouton 2001).[6]

One of the most ardent critics of the Royal Proclamation and the Quebec Act was none other than Thomas Jefferson, who argued that London had no right to make land policy for the colonies. This was the right, so thought Jefferson, of colonial councils. Friedenberg noted what Jefferson's position entailed in reality: "In effect, Jefferson was stating that the British government was unjust in its efforts to protect the Indians; but speculative grants and purchases of land based on wiping out the Indians were just. It was tyranny to protect the rights of the weak who had occupied the land for untold generations; it was liberty to seize this land, kill the natives, and then resell it at a profit to other white men who in many cases would then enslave blacks to farm it" (1992, 150–52).

Moreover, once the war broke out, Britain promised freedom to any slave or indentured servant who joined the British cause against the treasonous Americans. This proclamation further infuriated colonial leadership seeking independence and pushed many more moderates who benefited from slavery into the war on the patriot side (Bouton 2001, 669). Indeed, as Morris has pointed out, '[S]omebody might make out a plausible argument that the British fought to free the slaves and the Americans fought to keep them enslaved' (1962, 19). So while revolutionary patriots were denouncing imperial legislation as forms of tyranny and oppression, they fought hard to safeguard and defend their own private tyrannies over the slave population—the essential energy source of their wealth. The Constitution of the United States (1789), albeit tacitly, would be integral to securing the slave energy system by protecting racial domination and the ownership of other human beings.[7]

Like the revolution itself, the historiography on the making of the Constitution of the United States and the original intentions of its framers is vast and cannot be explored in any considerable detail here. However, whatever the scholarly disagreements, it is fairly clear that Federalists thought that the Articles of Confederation were too weak to protect property and advance their interests in the differential accumulation of money. Although colonial leadership no longer had to worry about Britain after the Treaty of Paris (1783), they faced obstacles at home. The chief fear was that the people had become too involved in their own governance and often initiated legislation that harmed the minority of affluent property owners. Debt and tax relief, as well as emissions of paper money, threatened the accumulation strategies of wealthy citizens. And where state legislatures were not responsive to popular demands for relief, distressed citizens often took up arms in open revolt, regularly justifying their resistance by appealing to the language of liberation in circulation during the revolutionary period (Holton 2005a, 2005b, 2005c; Wood 1969). The danger to elite rule and accumulation no longer emanated from a foreign Parliament and its king, but from the contradiction between the radically unequal distribution of property and the threat of majoritarian democracy. In the early United States, 10 percent of the population owned

50 percent of all colonial net worth (Huston 1993, 1093 and note 29). It was also well understood by elites that 'the most common and durable source of factions has been the various and unequal distribution of property' and that 'those who hold and those who are without property have ever formed distinct interests in society' (Madison in Woll 1996, 187). Moreover, because Madison (and likely all the framers) understood land to be the primary source of wealth, they also feared that in the future, as the population expanded, there would be threats to property. At the secretive Federal Convention whose records were not made public until 1840, Madison put it thus: "In future times a great majority of the people will not only be without landed, but any other sort of, property. These will either combine under the influence of their common situation; in which case, the rights of property and the public liberty, will not be secure. . . . How is this danger to be guarded against on republican principles? How is the danger in all cases of interested coalitions to oppress the minority [of property holders] to be guarded against?"[8]

This passage goes a long way in suggesting that the primary purpose of the Constitution was to safeguard the unequal distribution of property in perpetuity. Thus, the point was never to elevate and facilitate democracy, but to frustrate and channel it. The main way this was accomplished was by instituting a federal government because it was reasoned, contrary to the political theory of the time, that property could best be protected by enlarging the sphere of governance. As Holton (2005b) has suggested, Madison's strategy was to divide the population by extending the scale of government so that the minority, who could more effectively organize around issues and candidates, could rule it. In this way, Madison argued, 'a rage for paper money, for an abolition of debts, for an equal division of property, or for any other improper or wicked project' would be frustrated by the electoral system spanning the entire voting population.[9] Moreover, creating an aristocratic/plutocratic Senate to check the House of Representatives, as well as the creation of the Electoral College, the presidential veto and judicial review, all worked to frustrate majority rule and the popular will of the people. In fact, their design worked so well that two modern researchers found that average US citizens have virtually zero influence on policy and that policy reflects the will of business elites and organized interests (Gilens and Page 2014). Indeed, US historian Vernon Parrington's argument that 'the drift toward plutocracy was not a drift away from the spirit of the Constitution, but an inevitable unfolding from its premises' has considerable merit, particularly given the striking inequality and social disparities in the United States today (cited in Morgan 1957, 10). Indeed, the top 5 percent of the population in the United States today owns 63 percent of all net worth whereas the bottom 80 percent owns a meagre 11 percent.[10]

Moreover, under the Constitution, Americans lost the right to forceful rebellion against unjust rule and the organized means of violence became increasingly concentrated in the hands of the federal government. This was not so much inspired by fears of foreign invasion but by domestic insurrection against the exercise of government. The wealthy had direct experience of Shay's Rebellion calling for debt and tax relief and wanted to ensure they were powerful enough to put down any future rebellions (Engels 2005).[11] An organized federal force would also be used to violently expropriate certain native tribes of their land, although advances were cautious in the early years of the republic for fear of provoking more war with the Indians and thus more debt (Frymer 2014).

Indeed, one of the main reasons, if not *the* reason, for establishing a federal government was that the revolutionaries had incurred a massive debt to free themselves from imperial rule. Some of this debt was owed to foreigners, but most of it was owed to wealthy patriots (Davies 2002, 467). If the colonists were aggravated about paying what were essentially meagre taxes to London before the revolution, they found themselves riddled with taxation after the war.[12] Creditors wanted to be repaid for the revolutionary violence they successfully financed. States tried their best to collect taxes from their populations, but economic depression, retrograde tax bureaucracies (particularly in the South) and a scarcity of specie after the war meant that many could not afford to pay (Einhorn 2006). In this situation, creditors worried about being repaid with interest and even soldiers who had been paid in certificates or warrants sold their debt instruments to others who bought them at steep discounts, often in fraudulent ways—that is, if Jefferson can be believed (Sawvel 1903, 30). Moreover, elites like Alexander Hamilton understood that a new nation would need to borrow money and this meant establishing sound and trustworthy public credit.

Once the Constitution was ratified, Hamilton wasted little time enacting a financial plan to ensure debts could be serviced at face value, not to the original holders of the debt instruments but to its present holders, those who bought the instruments from others who were more desperate for actual money. As secretary of the treasury, Hamilton not only federalized state war debts—thereby relieving state legislatures of instituting harsh tax policies due to the war—but also decided that the federal government would pay the full face value of the debt. Those who bought debt instruments at steep discounts before Hamilton's plan was enacted stood to make incredible financial gains. To help service this debt, Hamilton introduced a regressive, yet not very controversial, tax on imports, particularly on wine, spirits, tea and coffee. He also institutionalized a central bank, the Bank of the United States, on the model of the Bank of England to help secure state credit and introduced a national currency based on the US silver dollar, though in practice other currencies remained in use (Cowen 2000; Edling and Kaplanoff 2004; Konings 2011, 28; Sylla 1998).[13] The majority of the Bank of the United States was privately capitalized, with the federal government only paying one-fifth of the bank's overall capitalization (Rothbard 2002, 68ff).

Perhaps even more important than the tariff and the bank in helping to service the federal debt were the frontier lands not yet in the new country's possession. The prospects of a new government being able to capitalize on western territories also convinced many foreign investors to help finance the war against Britain (Jensen 1958, 384–85). As Williams explains: "Not only were there speculators in the south, but the southern gentry saw the west as a capital fund to retire the national debt, and hence a good share of their particular debt" (2011, 134). The revolution had nullified the Royal Proclamation that protected—at least on paper if not in fact—the Indians from losing more of their land and hunting grounds to speculators and farmer-settlers. It is true that the new government proceeded cautiously in acquiring more territory. But there can be little doubt that *the idea* of future expansion and future profit from expropriated land existed in the minds of those who capitalized the state's power to oversee the dispensation of land, violently or otherwise (Banner 2005, 127; Frymer 2014; Weaver 2014, 190ff). The Royal Proclamation and Quebec Act blocked one of the only major avenues for differential profit seeking: the speculation on real estate. The Revolutionary War had succeeded in its task of overturning both decrees, and by the 1840s, 'almost all actual settlers had to purchase land from private speculators and land companies at prices well above the federal minimum prices' (Post 2014, 374). As in Britain, the capitalization of state power opened the field for an expanding securities market and a benchmark for differential accumulation. By 1800, 295 chartered companies in banking, insurance, dock and road construction and mining were formed (Sobel 1965, 19). However, this was not yet the era of the limited liability corporation, mass share issuance and widespread absentee ownership. Before the Civil War and the discovery and exploitation of oil, most capitalist industrial production, as in Britain, was family owned or run by small partnerships that largely capitalized muscle and water power (Nye 1999, 103–4). Moreover, the slaveocracy in the South was extremely suspicious of Northern finance and when in control of Washington DC did their best to pay down the 'national' debt that had been accrued in war or the preparation for war.

But a debt-free capitalist state unmortgaged to private social forces was wishful thinking. It is true that in 1834 the national debt was liquidated under the presidency of Andrew Jackson. However, thereafter, the United States continued to accumulate a 'national' debt for war and the preparation for war (Brown 1989). The 'national' debt was owned by a minority of bondholders who took a private share of the federal government's revenues for their trouble. It would only be in the twentieth century when debt would be accumulated by massive spending on both warfare and welfare.[14] Thus, if in the late period of the age of efflorescences the ruling class of Britain embarked upon a slow and cautious form of wind-powered outward colonization from its ports with coal-fired weapons, once they gained their freedom from London, the newly independent government of the United States embarked upon its own form of colonization by capturing more and more land while simultaneously constitutionalizing and effectively capitalizing human slavery through the 'national' debt—that is, insofar as federal power helped prolong slavery and the slave trade (Frymer 2014; Knowles 2007; Morgan 2001; Ohline 1971).[15] Indeed, slavery did not decline after the revolution for colonial 'liberty', but expanded for fifty years after the revolution (Baptist 2014; Rothman 2005, iv). According to Blackburn, on the eve of the American Civil War (1861), slaves considered 'property' 'were worth more than all the machines, factories, wharves, railroads and farm buildings of the North' (2011, 108–9). To be sure, slavery was a valuable energy system for early dominant owners of the nascent republic. The contradictions inherent in this system, combined with the increasing use of coal, would lead to one of the bloodiest battles in human history as the main drivers and beneficiaries of slavery fought to maintain their differential source of money and power in the opening stages of US carbon capitalism.

SLAVERY, CAPITALISM AND THE CIVIL WAR

Just as explanations of the Revolutionary War and the making of the US Constitution remain contested, so, too, are debates on the origins of the Civil War (1861–1865) and its aftermath (McPherson 2007). My purpose in this section is not to introduce any new arguments on the precise origins of the conflict, but to understand the way in which this slaughter over property contributed to the development of carbon capitalism in the United States. On the eve of the Civil War about 4 million African Americans were held as chattel slaves by about 300,000 Southern whites and a minority of Northerners in the border states (e.g., Delaware and Maryland) (Beckert 2004, 1406; Blackburn 2011; Marx 186). After the revolution, slavery had only very gradually eroded in the North, but it was never as present there as it was in the South where a mass labour force was needed to cultivate tobacco and rice. These plantations were the primary sources of white slave-master wealth, and they operated them for profit as their own private tyrannies. By the early nineteenth century, cotton became the new cash crop and created a frenzy for the expansion of slavery as a capitalized energy system in the South and Southwest (Berlin 1998, 360). As Beckert intones: "[T]he cotton industry was among the world's largest industries at midcentury, drawing on the labour of perhaps 20 million workers. Prior to 1861, most of the world supply of cotton had been produced by slaves on plantations in the American South and was spun into thread and woven into cloth by textile workers in Lancashire" (2004, 1405). Cotton, slavery and capital formed an unholy transatlantic trinity, all for the benefit of dominant owners in this nexus of cloth, forced labour and investment. Slavery in the United States had always been a territorially expansive system due to the fact that poor farming practices (e.g., monocropping) often degraded the soil within a generation or less. This was one of the major factors pushing Southern slave owners for more conquests in the South and Southwest: more energy from more fertile soil worked by slaves. More to the point, the accumulation of money and the social reproduction of 'gentlemanly' lifestyles demanded territorial expansion. In Marx's analysis, there were two additional reasons for expansion. First, adding more slave territories and then slave states to the Union was viewed as a political necessity to maintain control over the US Senate and thus have a veto in controlling the nation's legislation. The population in the North was growing rapidly and since the number of representatives of an individual state was contingent on the size of the population, the South would not be able to control the House of Representatives. But since each state was awarded two representatives, regardless of population size, the South could keep control of the Senate and thereby their power over national policy. From 1789 to 1858 there were either more slave states than free or their numbers were tied. By 1861, there were nineteen free states and fifteen slave states, shifting the balance of power in the Senate. As Einhorn states: "[W]e must stop beating around the bush on this issue. Slaveholding masters *did* rule much of the United States most of the time in this [pre-Civil War] period" (2006, 1; emphasis original). The third reason for expansion pointed out by Marx was that

> the number of actual slaveholders in the South of the Union does not amount to more than three hundred thousand, a narrow oligarchy that is confronted with many millions of so-called poor whites, whose numbers have been constantly growing through concentration of landed property and whose condition is only to be compared with that of the Roman plebeians in the period of Rome's extreme decline. Only by acquisition and the prospect of acquisition of new Territories, as well as by filibustering expeditions, is it possible to square the interests of these poor whites with those of the slaveholders, to give their restless thirst for action a harmless direction and to tame them with the prospect of one day becoming slaveholders themselves. (1861, 6)

Thus, the expansion of slavery as a form of differential accumulation was rooted in the need for a labour supply, the exhaustion of the soil, the control over the Senate, and the fact that land ownership was heavily concentrated, leading to a class of poor whites in desperate hope of land and their own slaves to work it. As hundreds of millions of acres were added to the continental United States by war, expropriation, or purchase, by mid-century, the question of slavery and the future political constitution of the United States intensified. The debate turned on whether these new territories (and later states) would be slave states or wage-labour states and as Marx rightly identified 'whether the twenty million free men of the North should submit any longer to an oligarchy of three hundred thousand slaveholders' (1861, 7). The Slave Power took the election of the Republican President Abraham Lincoln as a signal that their world was coming to an end. Lincoln, after all, had run on a platform that opposed the territorial expansion of slavery. To safeguard the slave mode of social reproduction and private accumulation, Southern leadership decided to secede from the Union. In 1861, the Confederate States of America was formed and its vice-president, Alexander H. Stephens, spelled out the fulcrum of the secession:

> The old confederation known as the United States, said Stephens, had been founded on the false idea that all men are created equal. The Confederacy, in contrast, "is founded upon exactly the opposite idea; its foundations are laid, its cornerstone rests, upon the great truth that the negro is not equal to the white man; that slavery, subordination to the superior race, is his natural and normal condition. This, our new Government, is the first, in the history of the world, based on this great physical, philosophical, and moral truth" (cited in McPherson 2007, 3).

The South was trying to perpetuate its racial reign over the kinetic energy and indeed the entire lives of African American slaves, mainly for profits now increasingly derived from cotton. As Baptist explains:

> Entrepreneurial enslavers moved more than 1 million enslaved people, by force, from the communities that survivors of the slave trade from Africa had built in the South and in the West to vast territories that were seized—also by force—from their Native American inhabitants. From 1783 at the end of the American Revolution to 1861, the number of slaves in the United States increased five times over, and all this expansion produced a powerful nation. For white enslavers were able to force enslaved African-American migrants to pick cotton faster and more efficiently than free people (2014, xxi).

After shots were fired on the federal outpost of Fort Sumter on 12 April 1861 in South Carolina, Lincoln had little choice but to mobilize for war. Although Lincoln was not a committed abolitionist and likely would have accepted existing slave states to maintain the Union, he eventually signed the Emancipation Proclamation in 1863 when the rebelling slaveocracy refused to rejoin the Union peacefully. The Thirteenth Amendment abolished slavery and involuntary servitude throughout the restored Union at the end of 1865. Human ownership and slave energy could no longer be capitalized for profit—at least not juridically. But the practices of domination would certainly continue. From the historical record, it seems likely that the antagonists did not anticipate the duration, expense and bloodshed that would be unleashed by the Civil War (McPherson 2007). Nor could the end of lawful slavery be foreseen. It is likely, as Baptist suggests, that the radicalization of the North during the war and the piercing voice of abolitionists opened up the possibility for the legal emancipation of the slaves (2014, xxvii). The Emancipation Proclamation (1863) could also be interpreted as a war tactic by the Lincoln administration because it threw the Slave Power into a frenzy, gained European (particularly British) support and encouraged more African Americans to enlist in the Union's campaign (McPherson 2007, 127). The war now had a greater cause than Lincoln's original goal of reestablishing the Union; it was now also about the fate of human ownership, freedom and the rights of men (McPherson 2007, 74). Yet there were also plenty of Northerners who shared the ideas of white racial supremacy and did not necessarily want the end of slavery (Berlin 1998, 358ff). As Baptist (2014) and others have shown, rather than two distinct 'modes of production', from the point of view of capitalization and accumulation, the system of slavery and so-called free labour were interdependent and intertwined at the level of the national and transatlantic economy. Far from destined to come to an end out of economic necessity so that capitalist social property relations could become the norm on the American continent, the capitalization of slavery was quite compatible with the forms of labour used in the North, such as waged and indentured work (Ashworth 2011; Foner 2011). What matters in the capital-as-power perspective—that is, from the point of view used in this study—is the class difference between owners of income-generating assets and nonowners of income-generating property, not specific forms of social labour or labour control, as important as they are in their own right.[16] For example, 'Northern merchants shipped the cotton, Northern financiers extended credit, Northern longshoremen handled Southern commerce (which went through New York City), Northern manufacturers turned slave-grown cotton into cloth' (Foner 2011, 95). Indeed, as Foner argues, 'the Lords of the Loom' had no conflict of interest with the 'Lords of the Lash' and there was little *economic* reason 'why the North could not continue to coexist for many years with a slave-economy in the South' (2011, 95). But if the war was *not* fundamentally about two *incompatible* economic systems, and clearly some Northern owners benefitted from Southern slavery, what is clear is that the war contributed to a form of capitalism based increasingly on carbon energy and perpetual war and the preparation for war. This is the dirty little secret of Jefferson's 'empire of liberty'.

THE DAWN OF AMERICAN CARBON CAPITALISM

Growing up alongside the low-energy agrarian society of the slave masters and their cotton revolution in the South was a Northern counterpart where different social forms of labour and energy prevailed. Just as in the South, there were owners and nonowners of income-generating property. This much Northern and Southern elites shared: they owned and capitalized property and drew an income from the right of ownership. The difference was that the majority of *nonowners* were not chattel slaves but largely wage labourers or indentured or domestic servants. Just above this class of dependents were those who owned small parcels of land, known as the yeomanry. This group of farmers was understood to have greater independence than those without access to the soil (Kulikoff 1992). As in Britain, two of the main factors that drew the yeomanry into the price system underpinning the capitalist mode of power were the need to make money for the purpose of paying tax or servicing debt for mortgages or farm implements. There is little doubt that some also engaged in market exchanges to earn money to purchase goods or services that could not be produced on the farm (Appleby 2001, 17; Wood 2002).[17] Still, there can be little doubt that considerable freedom from complete market dependence was something cherished by the yeomanry (Kulikoff 1992, 129ff). More exposed to market dependence were those living in the emerging towns. In a rapidly urbanizing antebellum North, those who had money invested in 'factories, mines, fleets of ships, and railroads', among other things, and established a more diversified economy than in the South (Nye 1999, 58). By the time of the Civil War, the North was also a far more energy-intensive economy using wood, charcoal, coal and water power for heat, pig iron, weapons manufacture and milling, just to list some of the main activities. Part of what explains this is the geography of coal and water in the North, as well as the climate and availability of timber relative to the South. As forests disappeared in the North, more and more energy-dense coal was exclusively relied on for heat and mechanical power so that by 1880, significant parts of the Northeast and Northwest relied on coal at a time when the Southeast and Southwest used timber for virtually all their nonanimate fuel (Williams 2006, 298). As Nye points out, '[T]he coal business grew by more than 65 percent each year, and often by more than 100 percent in a single year between 1820 (when 365 tons were sold) and 1834 (425,000 tons). The sudden abundance of inexpensive energy encouraged steam-powered manufacturing wherever it reached' and led to the greater concentration of production, as in Britain (Nye 1999, 76). Another part of the explanation was the lack of investment by Southern planters in any alternatives since they were largely content to have their slaves grow cash crops whose proceeds they could then use to purchase finished goods on the domestic and international market (Nye 1999, 54–5). Moreover, erected upon this store of energy emerged a number of Northern banks whose owners and directors could capitalize on the increasing capacity to do work provided by the steam engine in manufacturing and railroads in transportation. However, the use of steam engines in factory work did not outpace the waterwheel until 1875, and railroads 'emerged as the dominant transportation system in the middle of the nineteenth century. It clearly had technological momentum by 1850, and it did much to define the economy, landscape, and settlement patterns of the United States after 1840' (Nye 1999, 4 and 6). By the 1890s, dominant owners in the networks of finance and industry in New York were consolidated into a more self-conscious, propertied class with increasing access to energy from coal, oil and natural gas and a new way of organizing the institution of ownership (Beckert 2001; Nye 1999, 107). The energy, industrial and financial requirements of the Civil War and Reconstruction did much to advance the increase of capitalization, the individual appropriation of incredible amounts of wealth, the consolidation of ruling-class consciousness in the North, a more entrenched tax bureaucracy and the energy-intensive, limited liability corporation capitalized on Wall Street as the fount and matrix of the system. If the Civil War period can be conceptualized as a turning point in US history, it is in these dimensions where the transition should be sought (Pollack 2009, 1).[18]

First, the Civil War massively increased outstanding capitalization by snowballing the debt of the federal government and the span of taxation. The war was largely financed by monetizing state power, and like previous wars, gave investors the ability to capitalize state power with an expected 'return' on investment, thus deepening the capital market. As Brown reports: "[T]he nominal public debt rose from US$65 million at the beginning of the war to US$2.7 billion at its end" (1989, 7). This was the largest debt ever incurred in the United States, and it called for a vast tax apparatus to ensure it could be serviced:

> Congress took the lead in designing wartime fiscal policies, enacting the first federal income tax which reached a top rate of 10 percent, and an inheritance tax. The bulk of the revenue was provided by a massive array of excise taxes, both ad valorem and specific, that covered a large portion of market transactions, multiple taxes on manufactured products much like a turnover tax, stamp duties on a variety of transactions, and gross receipts taxes on such industries as railroads and utilities. These taxes required the creation of a whole new

administrative apparatus—the Office of Internal Revenue in the Treasury. (Brown 1989, 6)

Thus despite the greater fragmentation of banking and finance in the United States, the war debt–taxation–bureaucracy nexus, although alive and well at the dawn of the republic, really only came into its own after the massive war spending of the Civil War (Bensel 1990; Konings 2011, 9; Pollack 2009).[19] However, it would take the world wars of the twentieth century before the nexus was permanently woven into the nation's fabric.

Second, from 1875 to 1900, social reproduction in the United States was shifting from wood, water and animate power in some places to the greater use of coal. At the beginning of this period only about 1 quadrillion British thermal units (BTUs) were consumed, but by the end of our period that figure would hit 9 quadrillion BTUs (Anonymous 2011). To provide some sense of scale, consider that the highest level of coal consumption happened in the late 1990s and reached about 23 quadrillion BTUs. Thus, the 800 percent increase during our twenty-five-year period was historically unprecedented and incredibly significant given that from 1900 to the late 1990s there was only a 156 percent increase in coal consumption. The only thing comparable is the take-off of oil and natural gas consumption in the 1930s. Surely this era represented a 'big bang' in coal consumption and use.

Third, more coal energy coming online meant a greater capacity for industrial work and more opportunities for expanding industry, increasing the money supply through credit, and what would become a storm of capitalization. By the end of the nineteenth century, the clamor for foreign markets and the threat of overproduction (that is, production that cannot be realized for money) were unmistakable. 'America's factories', as Senator Beveridge exclaimed, 'are making more than the American people can use; American soil is producing more than they can consume'. By 1898, the State Department noted: "It seems to be conceded that every year we shall be confronted with an increasing surplus of manufactured goods for sale in foreign markets if American operatives and artisans are to be kept employed the year around. The enlargement of foreign consumption of the products of our mills and workshops has, therefore, become a serious problem of statesmanship as well as of commerce" (cited in Hybel 2012, 103).

In other words, the factories and mills of the United States had the capacity to produce beyond what the US market could pay for. The demand for foreign markets was equaled in the calls for a greater supply of money to circulate goods. Thanks largely to an era of 'free banking' begun in 1833, by 1860 there were 1,562 banks nationwide. As Davies notes, during this era 'there was a rising tide of money and credit supplied by a motley collection of banking institutions to meet the increasing demands of a nation where the population and, with occasional setbacks, the gross national product was growing at record pace' (2002, 480). In the post–Civil War period, banks continued to expand as energy consumption from coal (and later oil and natural gas) increased. Energy was being monetized and capitalized for private profit. From 1860 to 1921 a proliferation of banks reached 30,000 in number. Although they differed in size, locality and ability to extend credit, what all banks had in common was the differential power to create money as debt, either based on a reserve requirement or, for a small number of banks, no requirement at all (Davies 2002, 492–93).

But although banks proliferated during this era, there was still a rage for greater elasticity in the money supply. One factor seemed to threaten further economic expansion and threaten recessions: banks could only increase their note issues if they were backed by government securities. But since the federal government was not running 'a continuously increasing deficit' during the late nineteenth and early twentieth centuries, there were fewer government securities for increasing the note issues of banks. The creation of the Federal Reserve in 1913 as the nation's central bank, combined with the massive debt incurred by World War I, would never solve the scarcity of money problem, but did succeed in creating a more elastic supply.

Increasing energy use also led to an explosion in the capitalization of corporations during this period. Prior to 1890, few manufacturing firms were listed on the stock markets of the United States. Capitalization was largely confined to government bonds and transport and communications securities (Roy 1997, 3). The largest company in the world was the Pennsylvania Railroad, capitalized at a hundred million US dollars. In 1890, this figure alone dwarfed all the publically traded manufacturing companies, listed on exchanges at US$33 million. The following year, the capitalization of manufacturing corporations reached US$260 million. But if we consider all stocks and bonds traded on the public exchanges, we find an incredible increase, from almost US$1 billion dollars in 1898 to just over US$7 billion dollars by 1904—a 600 percent increase in five years. Manufacturing stocks accounted for US$6.8 billion of total capitalization by 1904. As Roy suggests, at this time the United States entered 'a new corporate order' supported by a wealth of fossil fuels (1997, 5). As we will investigate next, this was only the beginning. Although a number of characteristics could set the United States apart from the British Empire, by far the most important was the fact that it was the first Saudi Arabia, awash in petroleum 'from sea to shining sea'.

THE EMPIRE OF OIL

By the middle of the nineteenth century, the building blocks of the US petro-market civilization had largely been forged by slavery and wage labour, coal and water power, steel and rail and war and weapons. Railways, as in Britain, were the most heavily capitalized entities save the debts of governments, and connected major cities and towns with the hinterlands of agrarian production. Farming, too, was in the midst of change as new implements and machines were introduced to ease or accelerate production. For example, to produce forty bushels of corn in 1855 required about thirty-nine minutes of work, whereas in 1894 that figure was just fifteen minutes. A similar story can be told for cotton, barley, hay, oats, potatoes, rice and wheat (Nye 1999, 114). Banks continued to proliferate to supply producers and industrialists with much-needed credit, but overwhelming financial power was concentrated on Wall Street where by 1912 the Pujo Committee set up by Congress to investigate the 'money trust' found that a small coterie of men controlled a considerable part of the money and credit of the entire nation (Foster and Holleman 2010, 3). And controlling access to money and credit also meant that a small group of dominant owners came to exert their control over industrial firms in need of finance. From the mid-1800s, this was largely accomplished by exceptional violence to crush organized labour and a rash wave of mergers that largely centralized ownership and control over US industry and finance (Geisst 2000, 76; Hunter 1919; Josephson 1934; Perrow 2002; Roy 1999). This was the period of the Gilded Age in the United States, the 'robber barons', the accumulation of untold private fortunes and the rise of absentee ownership and the giant corporation (Veblen 1923). However, one firm would tower above all others and shape the terrain of US and global social reproduction in the twentieth century: Standard Oil.

Before the petroleum industry, the major source of illuminants and lubrication either came from animal and vegetable fats or better-quality whale oil. By the mid-nineteenth century, whales were on the verge of being hunted to extinction. In the United States alone, the whale fleet consisted of 736 sailing ships capitalized at US$3 million and producing revenues of US$8 million yearly (Black 2000, 16–17). Given the rapid pace of industrialization in the United States and the devastation of the whale population, interested parties began to search for a new source of illumination and lubrication. Petroleum had been known about on the continent for some time, but it was never found in sufficient quantities to be commodified and capitalized on a grand scale. This all changed in 1859 when Edwin Drake, working for the Seneca Oil Company, struck oil in Oil Creek Valley, Pennsylvania. For most of the 1860s, Oil Creek Valley was the world's largest producer of petroleum, netting over US$17 million for investors throughout its fourteen-year run (Black 2000, 5). It might as well be said that the great US oil rush began that day, and more and greater quantities of crude oil would be found in the ensuing decades. One man was poised to benefit immensely. He would become the world's first billionaire.

The story of Standard Oil and John D. Rockefeller has been told in detail elsewhere, so it bears little repeating (Tarbell 1904; Yergin 1991). Most readers will know that Standard Oil was incorporated in 1870, used various methods of sabotage to corner the market on petroleum products, and was eventually broken up by the US Supreme Court in 1911 under the Sherman Antitrust Act of 1890. However, what I believe is worth emphasizing is that Standard Oil, not to mention all emergent oil companies of the period, largely helped shape the terrain of global social reproduction by refining petroleum into kerosene for light.[20] Light extended the working day. Although oil did have other uses, illumination was the primary commodity upon which Standard Oil and its counterparts built their fortunes. But just as the monetization of petroleum was reaching a national and global fever pitch, an experiment in Lower Manhattan by Thomas Edison and his banker J. P. Morgan demonstrated the superiority of electric light in 1882 (Yergin 1991, 63). The proliferation of electric light meant the potential death knell for Standard Oil and the emergent oil industry.

One key development combined with the power of oil firms and state power saved the industry and propelled its capitalization forward. The key development was the more widespread adoption of internal combustion engines for motorized vehicles. The engines started to gain commercial appeal by the 1890s (Smil 1994, 132). But machines are little more than social artefacts outside of relations of property and power, and at the outset of the invention it was in no way clear that the oil industry would be saved by the gasoline-burning private automobile. Other than horses, three other options competed for commercial success in the private transportation market: the bicycle, steam-propelled cars and the electric car powered by batteries.

It is difficult to tell with absolute certainty why the internal combustion engine became the twentieth century's dominant prime mover of personal transportation in the United States and much of the world. Early on, the machines were dirty, noisy and difficult and sometimes dangerous to use. The electric car and bicycle had clear advantages—they were clean, silent and easy to use. The bicycle, of course, had disadvantages due to weather, but not so the electric car. Moreover, Henry Ford wanted to use alcohol as the primary fuel in his automobiles, particularly on farms where waste material can be converted into alcohol for fuel. However, 'it is alleged that his dream was shattered by John D. Rockefeller, who persuaded the Federal Government to put an exorbitant tax on alcohol, which made the cost of producing it prohibitive' (Henle 1980, 60). Ford also discussed with his longtime friend, Thomas Edison, the possibility of mass producing cars fueled by electrically charged batteries (Black 2006). That the gasoline-powered vehicle ultimately triumphed is often explained by its greater speed and power, as well as a marketing strategy that branded them as 'manly'. Though still foggy in considerable places, Black's research does suggest something more sinister. He argues that at the turn of the twentieth century, the battle for property and power among various capitalized monopolies—particularly the Lead Trust—essentially derailed alternatives to internal combustion engines. According to Yergin, by 1905, gasoline had 'defeated its competitors' and was now set to be the standard US fuel (1991, 64).

We will likely never know with any evidence-based accuracy why exactly gasoline-powered cars became normalized, but that the oil industry had an interest in making it so can hardly be in doubt. Save for in the rural countryside, electricity was putting an end to kerosene-derived light, and gasoline was largely thought to be a waste product, ejected into rivers and streams by refineries whose main commodity was kerosene. Gasoline-powered vehicles saved the oil industry, and in turn, the oil–automotive complex helped shape twentieth-century America, carbonizing and individualizing everyday life, first for wealthy and privileged white workers, and later for other groups and classes. In masterful detail, the story of the rise of what this study calls 'petro-market civilization' in the United States is best told by Huber (2013). So we will restrict ourselves to only a few comments here on how the oil–automotive complex helped shape and reshape social reproduction in the United States.

First, in 1900 there were only 8,000 registered vehicles in the country and most were steam driven, but by the end of World War II, 30 million gasoline-powered automobiles were on the road. According to the US Census of 2009, the number of automobiles, buses and trucks on the road stood at well over 246 million, or about 25 percent of the entire world's vehicle fleet of just over a billion vehicles (Nye 1999, 176; Painter 1986, 98; Tencer 2013; Yergin 1991, 64).[21] The domestic fleet of gasoline-powered vehicles is complemented by a kerosene-burning civilian air fleet of mainline carriers, regional carriers and cargo planes. The Federal Aviation Administration (FAA) reports that there were 6,156 such planes and a general aviation fleet of 202,865, and the Department of Transportation put the number of airports in 2013 at 19,453, up from just over 15,000 in 1980.[22] Furthermore,

> On any given day, more than 87,000 flights are in the skies in the United States. Only one-third are commercial carriers, like American, United or Southwest. On an average day, air traffic controllers handle 28,537 commercial flights (major and regional airlines), 27,178 general aviation flights (private planes), 24,548 air taxi flights (planes for hire), 5,260 military flights and 2,148 air cargo flights (Federal Express, UPS, etc.). At any given moment, roughly 5,000 planes are in the skies above the United States. In one year, controllers handle an average of 64 million takeoffs and landings.[23]

At present, there is no alternative to the use of petroleum in passenger or mainline aircraft. Thus, if space has been annihilated by time for a certain sector of the US population and its foreign visitors, the phenomenon is wholly powered by petroleum-derived fuels. In one sense, these developments have been absolutely crucial to the social formation and geography of the United States: its increasing suburbanization and the destruction or absence of public transit options. To be sure, urban centres use tremendous amounts of energy and 'consume about 75 per cent of global primary energy and emit between 50 and 60 per cent of the world's total greenhouse gases. This figure rises to approximately 80 per cent when the indirect emissions generated by urban inhabitants are included'.[24]

Urbanization and energy consumption are closely intertwined despite disparities in consumptive practices. But suburbanization contributes to greater energy intensity, given that it encourages what Campbell calls the 'distanciation of life' by creating an infrastructure that promotes the use of the private vehicle, not to mention a host of energy-intensive gadgets that proliferate each individual home (2005, 965). This 'American way of life' creates a vicious cycle where the social reproduction of everyday life is both petrolized and individualized, as well as gendered and racialized.[25] Paradoxically, this is a pattern of social reproduction that threatens to usher future generations of humans and the remaining species on the planet into a new and more dangerous world of climate change.

The private automobile did not create the suburbs. As an expression of carbon-dependent material life, they had begun as early as the 1850s for the well-to-do, and by 1895 one could witness their growth in the form of streetcar suburbs. But as Nye suggests, the automobile 'only changed the scale of suburbanization', not the cultural desire for mobility and a private home with a lawn (1999, 177). But although Nye is likely correct to suggest that the automobile did not cause the suburbs, it is equally true that the desire for a more privatized lifestyle away from the urban city was encouraged by government policy, developers and the willingness of banks to lend to the working class, as their increasing incomes could be capitalized in the form of mortgage debt. Creating such debt was also one of the main ways in which the money supply created as debt to banks could expand, all underwritten by cheap fossil fuels in the construction of urban and suburban environments, not to mention the industrial sector composed of growing petro-chemical firms and steel manufacturers (Campbell 2005, 965; Rowbotham 1998).

But carbon capitalism was neither wholly constructed nor completely dependent on the privatized motor vehicle, the propagation of aircrafts, or the urbanization and suburbanization of the US population. We know from our investigation of Britain that carbon capitalism from its inception had an intimate relationship with capitalization, militarization and war. Though it is difficult to say exactly how decisive this relationship has been for the development of the United States' unique petro-market civilization, that it was decisive in some considerable measure cannot be in doubt. Moreover, it would take considerable talent to miss the deep interconnections between the American way of life, internal stability and the foreign policy of its propertied, albeit nonpeered, aristocracy.

Like England after 1688 (later Great Britain in 1707), the propertied men of the United States were almost at constant war with continental social forces (native tribes, striking workers, Mexicans) or intervening militarily abroad (Blum 2004; Grimmett 2004). Given the sequencing in the timeline, wars and foreign interventions could hardly be seen as exceptions to a general rule of the search for markets, property and resources. But for the most part, the United States had avoided major conflagrations (though not conflicts) from the Civil War until World War I. World War I would lead to the beginning of a sea change in international leadership and world order that would only become a *fait accompli* after the devastation of World War II. World War I would make the United States a 'financial superpower' and extend carbon capitalism (Silber 2008, 2009; Sobel 1965, 220).

There are many intrigues about World War I, but since we are primarily concerned with the interrelations between energy, capitalization and social reproduction, we can restrict our focus to the effects the war had on capitalization, the use of oil and social reproduction in the United States. When the war began on 28 July 1914, the United States claimed neutrality and there was a strong pacific and isolationist mood among a considerable portion of the population. President Wilson had promised to keep the United States out of the war. However, this did not stop businessmen in the United States from financing the war, nor the owners of arms factories from producing war materiel that could be sold and shipped to Britain and France. In fact, the majority of the dollars borrowed throughout the war were to finance arms and oil purchases from US factories and refineries. Indeed, the export of oil accounted for 25 percent of total US production, and altogether the United States supplied the Allies with 80 percent of their petroleum requirements (Yergin 1991, 162).

The United States, in serious recession on the eve of the war, pulled out of recession as the war dragged on, creating greater demand for energy and materials as the open slaughter continued on the battlefields of Europe. As men were being gassed and blown to bits, other men in Britain, France and the United States were playing around with balance sheets to finance the war. By 1917, the production for war and oil sales moved the United States from a position of net debtor to net creditor (Sobel 1965, 221). But its chief debtors,

France, Britain and Italy, could not gain a decisive victory, and US financiers wondered what would happen to the debts they were owed if they lost the war (Engdahl 2004, 50ff; Michie 2006, 168) Furthermore, a key ally, Russia, was being exhausted on the Eastern Front and there were mounting agitations within Petrograd (now St. Petersburg) to end the war. Lenin's revolutionary government eventually called a cease-fire by the end of 1917. By April 1917, the United States joined the war on the side of Britain and France. The primary reason for entering the war is debatable, but there seems to be historical consensus that Germany's declaration of unrestricted submarine warfare against US merchant ships, the Zimmerman telegram (which proposed a German alliance with Mexico in the event of the United States entering the war), and the question of what would happen to the debts owed to Wall Street if the Allies lost were all high on the agenda. The total debt owed to financiers of the United States by the Allied powers was estimated at US$12.5 billion (Engdahl 2004, 56). But the real story was the increase in the national debts of the world. For example, the national debt of the United States went from US$1.3 billion in 1917 to nearly US$27 billion by 1919 (Sobel 1965, 27). The British national debt had increased by 924 percent from the start of the war to 1918 when it stood at the enormous sum of £7.4 billion. These debts were staggering for the time, but globally the war had caused a revolution in differential capitalization as J. P. Morgan & Co. partner Thomas W. Lamont noticed: "[A]s a result of the four years of war and global devastation, 'the national debts of the world have increased by $210,000,000,000 or about 475 per cent in the last six years, and as a natural consequence, the variety of government bonds and the number of investors in them have been greatly multiplied" (cited in Engdahl 2004, 55).

Part of this debt was made up of the US$33 billion demanded of Germany for war reparations, a key trigger for the next major European war, as Keynes rightly predicted (Engdahl 2004, 56). But although this massive increase in privately owned capitalization is crucial for understanding the rise of carbon capitalism in the United States, it is important to appreciate that it was primarily the energy derived from oil that was being monetized. Lord Curzon summed it up best: "The Allies were carried to victory on a flood of oil... With the commencement of the war, oil and its products began to rank as among the principal agents by which they [the Allied forces] would conduct, and by which they could win it. Without oil, how could they have procured the mobility of the fleet, the transport of their troops, or the manufacture of several explosives?" (cited in Engdahl 2004, 39; see also Yergin 1991, 167).

The expansion of capitalization, mainly through bank loans and the increase in the national debt, but also the growth on the New York Stock Exchange after its brief closure at the start of the war, was intimately tied to the consumption of oil, and oil to war (Silber 2008; Sobel 1965, 207ff). During the conflict, the United States produced 67 percent of the world's oil output and increased production from 266 million barrels in 1914 to 335 million barrels by 1917 (Yergin 1991, 162). Curzon's flood of oil was red, white and blue and for the few, green all over.[26]

THE GEOPOLITICAL CALCULUS AFTER WORLD WAR II

Although we do not have space to discuss it at length here, a similar story of rising national debt, increasing bank loans and accelerating capitalization occurred during World War II. 'The gross debt of the United States rose from $48 billion in 1941 to almost $260 billion by 1945' (Geisst 2000, 266). And despite government regulations after the stock market crash of 1929, the value of listed stocks on the New York Stock Exchange was US$55 billion, or an increase of 633 percent since the eve of World War I (Michie 2006, 213; Roy 1997, 5). By 1950, the consumption of oil and natural gas surpassed the consumption of coal in the United States for the first time, signaling a reliance on hydrocarbons as the dominant source of energy in the economy (Anonymous 2011). Despite constant worries about supply since at least the 1920s, the pursuit of differential accumulation by oil executives, industry and real estate developers further entrenched forms of social reproduction premised upon the availability of ever greater amounts of energy, particularly from oil. Through a range of policies from subsidies, the sale of public land, the public finance of roads and the procurement of war materiel, successive governments also promoted the carbonization of the economy and everyday life. Much of this energy would be wasted in war and the preparation for further wars during the Cold War with the Soviet Union. In fact, the interstate road network built over ten years and costing about US$21 billion was ordered in 1956 and, in part, justified on the basis of protecting the US population from a Soviet nuclear attack. The automobile industry's lobby group, the National Highway Users Conference, reasoned that since most citizens now lived in major cities, they would need to be able to flee in their cars to the hinterland in the event of a thermonuclear war (Rutledge 2005).

Although nuclear war did not take place, the policy did leave the United States with the most extensive road system in history and facilitated 'automobility' and continental trade that would later be transformed by the container and trucking industries. All this despite the knowledge that oil was a nonrenewable resource and that the United States would eventually exhaust its own domestic supplies. This was known as early as 1921 when petroleum engineer J. O. Lewis exclaimed that 'the desired supplies of petroleum in future cannot be obtained from our own wells. . . . This constitutes America's Petroleum Problem—how are we to obtain the necessary petroleum products that our growth be not stunted' (Lewis 1921, 375).

Growth and accumulation are the overarching logics, not the social reproduction of decent livelihoods for the entire political community. Moreover, the production of conventional oil peaked in the early 1970s, making the United States even more reliant on foreign oil than it had been previously (Heinberg 2003). To reproduce its high-energy–intensive lifestyles for the affluent, it now imports 3.5 billion of the 6.89 billion barrels of oil the country consumed in 2013.[27] With few exceptions, the solution of oil consumption in the United States was never to abandon oil, embrace alternative and renewable forms of energy, or promote large-scale conservation. The primary solution to the 'petroleum problem' was to find and acquire more of it, thereby further entrenching an unsustainable, hierarchical and environmentally disastrous form of social reproduction at home while at the same time encouraging more wars and greater militarization abroad (we discuss this in more detail in the next chapter). By 1961 Eisenhower was warning his fellow citizens about the undue influence of the military–industrial complex on US society, and scholars started to notice that whatever the rhetoric of its political leaders about peace and freedom, the United States had developed a permanent war economy (Custers 2010; Duncan and Coyne 2013; Melman 1970, 1974, 1997).[28] To justify this spending, which most citizens regard as highly wasteful, constant threats were conjured up—a trend stretching back to the origins of European settlement and one that Hofstadter called the 'paranoid style' of US politics (Campbell 1992; Hofstadter 1964). Carbon capitalism not only depends on the Pentagon and oil, but through the US military, protects the owners of these carbon energy assets. In effect, then, a publically funded armed force not only protects the assets of the dominant owners of oil and gas companies, but also the beneficiaries of the state-run firms of Organization for Petroleum Exporting Countries (OPEC).

There is little doubt that oil and certain other forms of energy can indeed improve livelihoods and the general standard of living. But this was never the primary aim or drive of carbon capitalists. Combined with an advertising and marketing juggernaut, the design was to perpetuate the differential accumulation of money for dominant owners while keeping the nonowning working class complacent. This logic is encoded in corporate culture where the executives running companies compete to increase their differential profit and therefore provide the basis for rising dividends and share prices. Carbon capitalism is not an economy of livelihood, but of differential property and capitalization. It is ultimately an economy where dominant owners are monetizing the destruction of the biosphere for differential momentary benefits and symbolic power (Di Muzio 2015b). How long that moment can last and some of its major consequences are the subjects of our next two chapters.

NOTES

1. http://www.eia.gov/countries/index.cfm?view=consumption
2. My calculations using SIPRI's Military Expenditures Database.
3. Data for stock market capitalization are from the World Federation of Exchanges, November report (http://www.world-exchanges.org/statistics/monthly-reports) and the date for the US national debt is from *The Economist*'s debt clock.
4. Consider, for example, Einhorn's description of the Virginia Company:

> The Virginia Company, which was chartered by James I in 1606, chose these settlers to pursue what turned out to be a series of harebrained schemes to bring wealth back to England: goldsmiths and refiners to process gold that did not exist, glassmakers, silk dressers, perfumers, apothecaries, and a huge oversupply of 'gentlemen', who paid their own way to Virginia but had no plans to engage in manual labor. (2006, 32)

5. Dr. Samuel Johnson, *Taxation No Tyranny*. 1775 online: http://www.samuel-johnson.com/tnt.html.
6. We have focused on land here, but there were, of course, some additional reasons why the colonists decided to take up arms against their mother country, such as the high tariffs resulting from the Navigation Acts, the strictures of the Currency Act of 1764 and the Intolerable Acts designed to punish Massachusetts for the Boston Tea Party (Curtis 2014, 455ff; Greene and Jellison 1961, 486; Sosin 1964, 175).
7. In 1801, after the Haitian Revolution, a different written constitution was born. Article 3, Title 2, states that '[t]here can be no slaves on this territory; servitude has been forever abolished'. To my knowledge, this is the only written constitution in history that forbids slavery—a document written in the same epoch when the Americans effectively constitutionalized slavery (https://www.marxists.org/history/haiti/1801/constitution.htm).
8. James Madison, Records of the Federal Convention, chapter 16, document 16 (http://press-pubs.uchicago.edu/founders/documents/v1ch16s16.html) and James Madison, The Constitutional Convention Debates (http://grid.let.rug.nl/~usa/D/1776-1800/federalist/anti24.htm).
9. James Madison, *Federalist 10* (http://www.constitution.org/fed/federa10.htm).
10. http://www2.ucsc.edu/whorulesamerica/power/wealth.html.
11. See also Alexander Hamilton's *Federalist 21* and *23*.
12. It should be recalled that the same situation prevailed in England after 1688. Taxes skyrocketed to finance wars (Brewer 1989).
13. Under the republican hegemony of Jefferson, the bank was abolished.
14. Local and state debt, however, were accumulated for large infrastructural projects, particularly but not exclusively, in cities. Capitalizing local and state power to collect taxes to pay for these projects with interest to creditors is certainly not

negligible, but the federal debt of the United States has far outstripped state and local debt. The ratio at the time of this writing (January 2015) was 6:1. See http://www.us-governmentdebt.us/. See Brown (1989) for early republic state and local debt figures.

15. Note that the importation of slaves was officially banned in 1808 by Congress and this was recognized in the Constitution. This did not stop the internal trade from developing and expanding, nor did it stop the illegal importation of slaves. Children of slaves were reared as slaves and thought of as tradable property. In this way, many families were broken up by slave masters who used slaves to pay for land or other goods (Baptist 2014).

16. What matters to capitalists is increasing profit and capitalization, not necessarily what source of labour helps them accomplish this; here I would argue they are relatively agnostic. If they can get away with slavery, they will; if not, other forms of labour control become possible. However, given the ubiquity of wage labour, one might make the plausible case that it is a more effective system of labour control for dominant owners.

17. For Wood (2002) this is the market as opportunity rather than the market as compulsion.

18. The available historical statistics suggest that the Civil War may have actually retarded the rate of industrial growth that had already gathered pace from the early decades of the 1800s when more coal was coming online (Cochrane 1961). Harvey (2012) also convincingly argues that the Civil War was not the first 'modern' war and that it produced no significant military innovations that were not already available in Europe.

19. With proof that all things are relative, Harvey (2012) also notes how the Civil War debt paled in comparison to the finance needed to finally defeat Napoleon.

20. Rockefeller would later add commercial banking to his interests (Rothbard 2002, 188). After a series of mergers, the bank is now known as JP Morgan and Chase.

21. http://www.census.gov/compendia/statab/2012/tables/12s1096.pdf.

22. http://www.rita.dot.gov/bts/sites/rita.dot.gov.bts/files/publications/national_transportation_statistics/html/table_01_03.html.

23. http://1lawflying.wordpress.com/2008/11/04/how-many-flights-per-day-do-air-traffic-controllers-handle-in-the-united-states/.

24. http://unhabitat.org/urban-themes/energy/.

25. Petrolization and petrolized are terms used by Matt Dow in personal correspondence.

26. It should also be noted that Shell played a key role in helping Britain prosecute the war (Yergin 1991, 159).

27. http://www.eia.gov/tools/faqs/faq.cfm?id=33&t=6 and http://www.eia.gov/dnav/pet/pet_move_impcus_a2_nus_ep00_im0_mbbl_a.htm.

28. This is sometimes referred to as military Keynesianism—the notion that state spending on war and defence stimulates the economy.

5

Global Carbon Capitalism

It should be clear from our analysis so far that the construction of a highly uneven and hierarchical petro-market civilization on a global scale was never dislocated from the apparatuses of force and the near-continual application of violence based on constant technological innovation in the leading belligerent states. As Luxemburg pointed out long ago, '[F]orce is the only solution open to capital: the accumulation of capital, seen as a historical process, employs force as permanent weapon, not only at its genesis but further down to the present day' (Luxemburg 1913, 371). As carbon capitalism became more institutionalized and organized, this was *always* the result of a fusion of state fiscal and growing corporate power. Where greater energy could be harnessed, more elaborate apparatuses of violence could be constructed and more domination over populations and resources could be applied against resistance internationally. In this sense, thinking of a 'state' sphere and a separate 'economic' sphere appears very odd in the light of history, where we find them everywhere inextricably interwoven.

I have also tried to show how these developments have chiefly benefited dominant owners, or what we are now calling the 1 percent (Di Muzio 2015a). This is not to say that benefits were not gained by the lower classes as carbon capitalism developed. Surely it cannot be denied that in certain parts of the world 'living standards' and opportunities and life chances were enhanced for a considerable size of the global population with access to greater energy flows and stores. But whatever these achievements, particularly in the global north, we can argue that they must be viewed as the result of struggles for recognition and rights within the historical context of rising energy use and greater global militarization (Mitchell 2011; Nikiforuk 2012). In addition, we could make the reasonable claim that these developments should also be understood within an ongoing intraclass and interclass war for differential accumulation on a far more global scale than was possible only centuries ago. This is not to deny the poor, working class, or disenfranchised masses of the world their agency. Dominant owners have always come up against opposition and resistance to the exercise of their will to power over others and the control of the natural world (Foucault 1982; Gill 2008). But the chief difference between the 1 percent and the rest is this: their wills are almost always backed by decisive violent and institutional force and these forces are, more often than not, the result of uneven access to energy, money and political power.

Where we find hierarchies of power and privilege and extreme divisions of social wealth appropriated by the few, we can be sure to find hierarchical forms of agency and justifications for this hierarchy. Whatever the weapons of the weak, it is simply not true to say that workers or nonowners of society's income-generating assets have the same access to resources, political power and weaponry as their masters. Even where there is a semblance of civil peace, we should not delude ourselves into thinking that this is a 'normal' state of affairs, never to be challenged or changed for all eternity. In the coming decades, global climate disasters and rising energy prices will almost certainly challenge existing social relations of power by contributing to greater civil strife as life chances and opportunities radically diverge from expectations. In the future, the state of emergency may not be exceptional but globalized and perpetual (Agamben 1998). Indeed, two key innovations in our own time should help us remain sceptical: (1) the militarization of police forces in North America and Europe; and (2) the development of greater technological and surveillance capacities of the state combined with the development of a range of 'less' lethal weaponry for crowd control, such as the Active Denial System and the Mob Excess Deterrent Using Silent Audio (MEDUSA).[1] Given carbon capitalism's emergence in war and ruling-class power, should we be surprised to find that the trend continues today? And if it does, should we not be keenly aware of the shifting energy landscape and the international power dynamics that are shaping and reshaping the social reproduction of a twenty-first-century world order? These broad questions, the continuities, transformations and likely trajectories of global carbon capitalism are the subjects of this penultimate chapter. I propose an investigation along three lines. First, I will consider the new energy environment and what the *Financial Times* has called the 'new seven sisters'—an updated list of state-owned oil and gas companies now in control of most of the world's oil and gas reserves and their extraction, production and sale. We will also consider how the four BRIC countries (Brazil, Russia, India and China) have fundamentally altered the world energy order by seeking to build more energy-intensive environments of their own, also animated by the logic of differential accumulation. With this landscape charted, if not fully explored, I then discuss the role of petrodollars, conflict and differential redistribution as key aspects of modern carbon capitalism. Here, I will argue that adequately conceptualizing the links between what could be called a finance–energy–armament nexus since at least the 1970s helps to shed considerable light on why world order continues to be shaped and reshaped as a global petro-market civilization when virtually all major indicators, such as the devastation of the biosphere and the nonrenewable nature of fossil fuels, would suggest a different course of action and the search for alternative forms of social reproduction. Last, since the only direct competitor with the fossil fuel industry is the renewable energy industry (and, for some, nuclear), in the final section of this chapter I will investigate some key trends in the renewable energy industry and question whether it is likely to overtake fossil fuels as a leading rather than auxiliary source of energy production and consumption. There are a number of ways to do this, but my main concern will stress a comparative focus on differential capitalization in both industries.

As the reader will see from the empirical evidence, there is little hope for a direct, rapid, or immediate transition to a renewable post-carbon era that can sustain current, let alone expanding, consumption practices. I do not say this because I have a desire for future catastrophe, but given the evidence, I have no desire for future fantasies either. Our feet will remain firmly on the ground with the evidence.

THE NEW SEVEN SISTERS AND THE NEW ENERGY LANDSCAPE

At the dawn of the twenty-first century, it was clear to most observers that the energy landscape had changed from the one shaped by the traditional international oil companies in the twentieth century. To be sure, this change was rooted in the 'resource nationalisms' of the twentieth century, first begun by Mexico in the 1930s. This change did not happen overnight, of course, but was a gradual process, the compound results of which are now relatively clear. At the beginning of what we have called the 'little oil age' within the broader age of carbon energy, seven major international oil companies, five based in the United States and two within the United Kingdom, controlled in one way or another most of the world's oil. Together they were not only used as instruments of Anglo-American foreign policy, but also shaped a hierarchical, gendered and racialized world whose social reproduction increasingly came to rest on the extraction, production and consumption of nonrenewable energy (Nore and Turner 1980; Vitalis 2007). The Italian businessman, Enrico Mattei, called them the '*sette sorelle*', or 'seven sisters' in English. The siblings were composed of Anglo-Persian Oil Company (now BP); Gulf Oil, Standard Oil of California (SoCal) and Texaco (now Chevron); Royal Dutch

Shell; Standard Oil of New Jersey (Esso) and Standard Oil Company of New York (Socony) (now ExxonMobil). As a group, and before their mergers in later decades, they actively blocked access to international oil when it was not in their interests to work with independent firms. In this quest for power, they were largely assisted by their governments (Mitchell 2011). However, despite attempts to thwart nationalization drives through assassinations and coups and the support of human rights–abusing autocratic rulers, the companies were neither all powerful nor the violence and elite consensus building of Anglo-American governments infallible. Nationalizations did take place, and with increasing frequency, so that the oil majors only own 3 percent of global oil reserves as assets on their balance sheets and only produce 10 percent of the world's oil. What's more, no publically listed oil and gas firm has been able to find wells that can produce more than 1 million barrels a day since 2000 when the 'Kashagan field became the biggest find in 30 years'.[2] Because oil companies are largely capitalized on the basis of their reserves and ability to find more oil and gas resources to book on their balance sheets (what is called the reserves replacement ratio), this is likely to be a worrisome trend in the future as reserves start to dwindle and the likelihood of finding giant oil fields becomes scarcer. There are only two avenues outside of this scenario: the first is to merge with another firm so corporate executives can

Table 5.0.

Reserves by Country	
Venezuela	297.6
Saudi Arabia	267.9
Canada	173.1
Iran	154.6
Iraq	141.4
Kuwait	104
United Arab Emirates	97.8
Russia	80
Libya	48
Nigeria	37.2
United States	33.4
Kazakhstan	30
Qatar	25.4
China	23.7
Brazil	13.2
Algeria	12.2
Angola	10.5
Mexico	10.3

Source. Energy Information Agency, 2013.

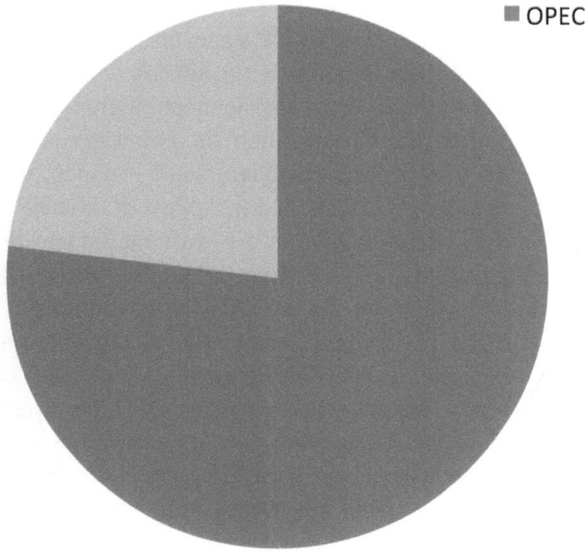

Figure 5.0. OPEC and non-OPEC Oil Reserves, 2013. Source: IEA

record the newly required reserves as assets on a single company's balance sheet, or second, as we will discuss in detail in the next section, hope for or actively encourage conflict in the Middle East to boost the price of existing reserves. Mapping global reserves in the new energy landscape helps us to see why. First, 117 countries have no reserves whatsoever, and a further 80 countries have less than 10 billion barrels of reserves. Second, as table 5.0 reveals, most reserves are now state owned.

Of the eighteen countries with a surplus of 10 billion barrels of oil in reserve, only seven are non–Organization for Petroleum Exporting Countries (OPEC) countries: Canada, China, the United States, Kazakhstan, Mexico, Russia and Brazil. They account for only 363 billion barrels of the 1.57 trillion barrels of world reserves represented graphically in figure 5.0. The bulk, 1.2 trillion, are owned and controlled by state-run oil and gas firms, mostly of the Middle East, at 851 billion barrels of oil for the region.

Because the overwhelming majority of reserves and production capacity has shifted from the international oil companies to the global south, the *Financial Times* has suggested an updated list of newly powerful companies outside the Organization for Economic Cooperation and Development (OECD) called the 'new seven sisters': Saudi Aramco, the National Iranian Oil Company, the Iraqi National Oil Company, PDVSA of Venezuela, PetroChina, Gazprom, Petrobas and Petronas. In consultation with industry experts, the *Financial Times* argues that these state-run firms are the most internationally 'influential energy companies outside the Organization for Economic Cooperation and Development'. To the best of our knowledge, Saudi Aramco, whose oil policies are dictated by the House of Saud, is the biggest of the new seven sisters, aptly called 'the world's central banker of oil—turning the taps on when there is a shortage of global supply, and off when prices are falling below its comfort level'.[3] In addition, although the earnings of oil companies often reach the mass media spotlight, the revenues and profits of the state-run companies are far larger.[4] For example, ExxonMobil's earnings were US$32.5 billion in fiscal year 2014, whereas the estimated earnings of Saudi Aramco for 2013 are just over US$227 billion.[5] Thus, the executive directors, combined with the policies of their governments, are managing and allocating massive sums of petrodollars and how they spend this money influences the constitution and reconstitution of world order. As we discuss in more detail in the next section, this will have potentially dramatic effects going forward, not least of which the increasing militarization of the Middle East and the consolidation of wealth and the ability to allocate it in fewer hands.

But the ownership of the production of reserves is not the only thing that has changed since the original seven sisters controlled the world oil order. Another shift is the increasing consumption of Brazil, India and China. This infectious quest to develop high-energy modes of social reproduction is leading toward the intensification of competition for the world's remaining resources, particularly for oil (Klare 2012).

Although some predict future resource wars for oil and gas and some scholars argue that resource wars are nothing new in the international system, what is clear is that energy consumption has been shifting to the BRIC and other emerging markets. According to *BP Energy Outlook 2013*, energy consumption by the four BRICs was a mere 23 percent of world consumption in 1990. This is estimated to increase to 42 percent by 2030. Moreover, the International Energy Agency (IEA) estimated that in mid-2013, a historical tipping point would be reached when for the first time non-OECD countries will consume more oil than the OECD.[6] Using data gathered by the Energy Information Agency, figures 5.1 and 5.2 track the growth in energy consumption of the BRIC and notes their percent change.

Future resource wars should not be ruled out, but what is for certain is that the energy consumption landscape has shifted in the twenty-first century as countries embark or continue to embark upon a path of development whose mode of social reproduction is ultimately environmentally ruinous and, over the long term, nonrenewable. The absurdity is that if these transitioning or

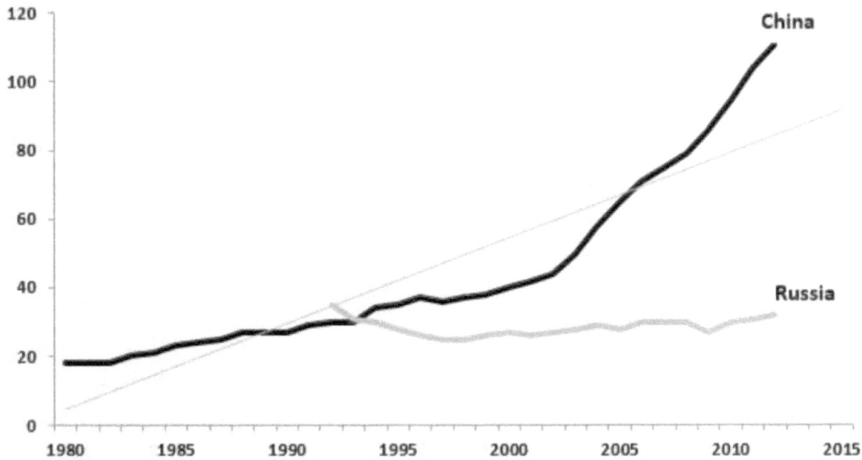

Figure 5.1. Total Primary Energy Consumption, China and Russia, 1980–2012, Quadrillion Btu. Source: EIA

developing countries continue to build energy-intensive urban and service infrastructures, mass private motorization and heavy industry, they may reach the pinnacle of their goal of achieving a petro-market civilization of mass consumption precisely at a point in time when competition for energy and increasing prices force them into making decisions about the future with little time for planned rational action. In these situations, there is little doubt that the prospects for violence could be great.

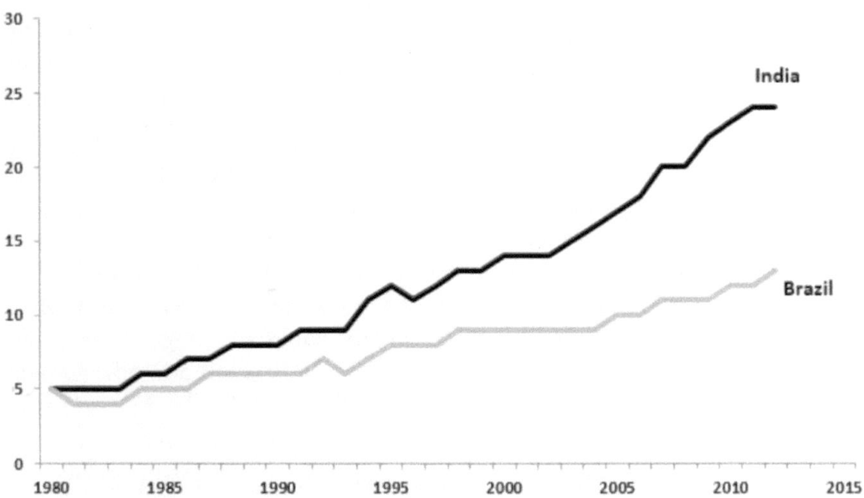

Figure 5.2. Total Primary Energy Consumption, India and Brazil, 1980–2012, Quadrillion Btu. Source: EIA

But we should not just be concerned with the potential for future resource wars if carbon capitalism is allowed to continue. If history is any indication, we are likely to witness more extreme forms of politics and domestic violence as people scramble to understand what is happening to their lives and look to answers from demagogues. Make no mistake: although scarcity has certainly been engineered by capitalists to boost their profits and the sabotage of production and social reproduction is an ongoing facet of differential accumulation, when oil starts to become radically more expensive, you can bet that more and more of the commodities energy-intensive consumers take for granted will become increasingly more scarce or unaffordable for the majority of humanity. There are very real dangers here that we will explore in greater detail in our concluding chapter. But with the new energy landscape sketched and with our knowledge of how ownership has transformed the world oil order, we now move to a critical analysis of one of the most overlooked and understudied phenomena of the global political economy: petrodollars, global redistribution and permanent war.

PETRODOLLARS, REDISTRIBUTION AND PERPETUAL WAR

The emergence of the Bretton Woods system at the end of World War II is a familiar story in international political economy (IPE). Before the war ended, a series of discussions were held in New Hampshire with the view of reestablishing an international monetary order and reconstructing Western Europe at the close of the war. Two major international financial institutions were created: (1) the International Bank for Reconstruction and Development, charged initially with assisting certain European countries to rebuild after the war and (2) the International Monetary Fund (IMF), tasked with facilitating and stabilizing international trade.[7] The IMF was to accomplish its task in two main ways. First, the institution was charged with alleviating what was thought to be the 'temporary' balance of payments problems suffered by those countries that imported more than they exported.[8] Second, the IMF was to maintain a system of fixed (but slightly adjustable) exchange rates where the currencies of participating countries were pegged to the US dollar and the US dollar, in turn, was pegged to gold at US$35 an ounce. In this scheme, countries earning US dollars could exchange dollar bills for gold. But since US dollars were needed to purchase oil and other goods denominated in dollars, as long as the United States maintained a trade surplus with the rest of the world, there would be a strong demand for dollars (Eichengreen 2008, 127). This system lasted only a short while when the private market for gold shot above US$35 dollars and US military spending escalated to finance its savagery in Indochina and Johnson's 'Great Society' program at home. As a consequence, nations holding US dollars as a reserve currency started to demand US gold instead of paper. Once the calculations were performed, however, US geopolitical strategists quickly understood that their gold reserves were likely to be depleted rather quickly, given the volume of US dollars outstanding. Nixon was advised to close the gold window and promptly did so in 1971, unilaterally throwing the Bretton Woods system into ruin and ushering 'the world economy onto a pure dollar standard' (Gowan 1999, 19–20). In this narrative, however, the role of petrodollars in backing up the new monetary order has often been overlooked (Clark 2005; Mitchell 2011, 110; Shipley 2007).

Because US companies were the first to develop the oil industry, the vast majority of the sale of petroleum has traditionally been in US dollars.[9] Moreover, US diplomacy with Saudi Arabia and the OPEC agreed that the world's largest reserves of oil would be sold in dollars (Spiro 1999, 110). Because any country that wants to industrialize or develop more energy-intensive forms of social reproduction must purchase oil if they do not have it within their own territory, this creates a significant demand for US dollars and is one of the principal reasons why the US dollar remains the most significant reserve currency in the world (Clark 2005). Given that the United States exhausted the first half of its supplies in war and automobility and the fact that the vast majority of oil lies outside the United States, there is no natural or even economic reason why oil should be priced *only* in dollars (Clark 2005).[10] Moreover, it would make a lot more financial sense for those nations who trade heavily with OPEC to purchase oil with their own currencies, thus eliminating currency risk involved in trading one's currency for US dollars to buy oil. There is, however, a political reason why oil continues to be denominated in dollars and that largely has to do with the Pentagon capitalism of the US government and its ability to project international military force. Understanding how this developed is crucial for understanding how US power shaped and reshaped a world order for dominant owners and the creation of a more universal petro-market civilization.

By the time the Nixon administration fretted about the future of the dollar–gold link, his strategic advisors understood that the dollar was the de facto world reserve currency. They also knew that there would likely be a continued high demand for US dollars given the size of the US securities market, the eurodollar market, financial innovation on Wall Street and the fact that a range of internationally traded commodities were denominated in US dollars (Konings 2011, 123). In fact, it was a senior economist from Standard Oil who wrote the most influential argument for the United States to abandon the gold standard (Mitchell 2011, 171). The most important commodity by value and volume was oil, followed by the demand for US weapons hardware and technology (Mitchell 2011, 111). This much is more or less agreed to in the scholarly literature, with the remainder of the story about the break with gold and the subsequent oil shocks considered controversial or largely ignored. Part of the reason for this controversy and neglect, I propose, is that virtually all the scholarly literature on the important developments after the dollar break with gold has failed to consult or cite a revealing article that appeared in the winter issue of *Foreign Policy* in 1976–1977 (Clark 2005; El-Gamal and Jaffe 2010; Eichengreen 2008; Engdahl 2004; Gowan 1999; Hudson 2005; Konings 2011; Nitzan and Bichler 2002; Panitch and Gindin 2012; Spiro 1999; *the exceptions are Parra 2004 and Mitchell 2011*). The author of the report, V. H. Oppenheim (1976–1977) examined State Department documents, congressional testimony and printed source material and interviewed policymakers of the period to assess how oil prices were suddenly increased when an esteemed oil expert and economist presented overwhelming evidence that supply could not be the issue. At the time, Adelman demonstrated that 'supply has not only *not* tightened, it has been getting easier' (1972–1973, 73, my emphasis; see also Nitzan and Bichler 2002, 230). He expressed disbelief at the fact that the State Department seemed to be encouraging the increase in oil prices well before the giant price shock that came after the Yom Kippur War of October 1973.

Viewed from his perspective of US nationalism (that is, that the government acts in the interests of all Americans), the State Department's action does appear rather peculiar. Why would the US government encourage collusion between OPEC and the oil companies and help them promote the idea that there were supply difficulties? After all, the oil companies through their industry lobby, the American Petroleum Institute, began a massive advertising blitz 'to familiarize Americans with the "energy gap"'—a gap that did not exist according to Adelman (Oppenheim 1976–1977, 29). Adelman was further concerned by discussions that OPEC had any real power to influence oil prices without the tacit acceptance of the US government. Moreover, he argued that the United States could have selected from a range of policies to persuade OPEC to keep its pricing in line with actual supply and demand, from freezing bank accounts to the threat of military action (Adelman 1972–1973, 79). However, if we shift our gaze from a methodological nationalism and start to see the world of power politics through the eyes of differential accumulation and the importance and influence of oil and armament firms and their owners, the US government's support for OPEC, the oil companies and price increases makes considerable sense. As stated earlier, the evidence found in Oppenheim's article as well as Adelman's and pronouncements by a former Saudi oil minister strongly suggest that the US government supported moves to increase oil prices.[11] Oil prices went from US$10.64 in 1970 to US$17 in 1973. By 1974, a barrel of oil cost US$53.94—a 407 percent increase from 1970.

The important question is why this spike in prices was encouraged and here a number of reasons have been addressed in the scholarly literature. But whichever reason or reasons we might find to be more important, they all had to do with taxing the global oil-consuming population for the *primary* benefit of the oil, arms and financial industry, largely of the United States but with considerable benefits for the City of London, given the centrality of eurodollars to British finance. First, the increase in price amplified the demand for dollars globally since more dollars were now needed to pay for oil. What this means is that countries in need of oil to 'develop' now needed to sell more goods on the international market to earn more dollars to pay for oil. Second, since OPEC could not absorb (or did not absorb) the windfall of petrodollars in their own financial institutions, governments, or industry, the petrodollars were parked in Anglo-American banks or government securities. In testimony to the Senate Subcommittee of Financial Markets, then Secretary of the Treasury William E. Simon (a former senior partner of Salomon Brothers, now absorbed into Citigroup) noted that OPEC countries made US$95 billion in 1974 alone. They spent US$35 billion of this money on consumption, with most of the remaining US$60 billion, or 66 percent, flowing into US banks and the eurocurrency market as bank deposits, US and UK government securities and commercial paper. Only a small amount of equity and real estate purchases could be traced, and Simon testified that 15 percent of where the money went 'we simply know nothing about' (US Senate 1975, 5). From the perspective of the commercial banks, this had the benefit of increasing bank deposits, which are liabilities on bank balance sheets, not assets. This forced the banks to find greater avenues for lending to generate profit that would help them service the interest on petrodoller deposits.

The third major beneficiaries were the dominant owners of the arms industry, as OPEC nations were now flush with surplus dollars and rushed to purchase an array of weaponry. Under the guise of fighting communism, the State Department cleared the way for the arms trade (Nitzan and Bichler 2002). From the mid-1970s arms sales were crucial for US armament firms as the Vietnam War was winding down and companies suffered the danger of excess capacity. Moreover, an internal study carried out by the Department of Defense demonstrated that foreign arms sales were 2.5 times more profitable than domestic sales (Nitzan and Bichler 2002, 214). With the Middle East in the lead, the 'Third World arms imports swelled from $6.2 billion in 1969 to $15.5 billion in 1978' (Stohl and Grillot 2009, 22). As Nitzan and Bichler observed, US foreign policy was largely 'bound up with private profit' (2002, 250). Fifth, increasing prices provided oil companies with super-profits that they could use to develop more difficult wells and where they were not self-financing through cash reserves, service their debts to the banks with greater ease. Production prices in the United States were escalating as old wells were being exhausted (Bina 1988; Rutledge 2005, 39). Windfall profits could help the majors expand into new fields. What made matters worse was a wave of nationalizations that stripped the major oil companies of their ownership over reserves. This put them in a subsidiary role as partners or advisors to the newly created state-run oil companies and forced them to find more reserves in riskier parts of the globe to book on their balance sheets. Their future earning capacity and capitalization depended on it. Presumably the oil companies needed a high price for oil to finance the North Sea development as well as Prudhoe Bay—both relatively inhospitable territories posing significant technical challenges for oil production. Some scholars argue that a final reason for using the oil price as a weapon against the global population of oil consumers was to weaken Japan and Europe whose companies were eating into the market share of US companies. Since Japan and Europe were more dependent on Middle Eastern oil than the United States, a ramped-up oil price could damage their industrial growth (Gowan 1999; Engdahl 2004; Mitchell 2011). Strategically intended or not, it is highly unlikely that the Nixon administration was unaware that mounting oil prices would cause severe strife in the Third World, creating the need for new loans to finance petroleum imports among other things. After all, the Nixon administration did intend to allow the private banks to recycle petrodollars, and Spiro (1999) documented how policymakers actively sabotaged the IMF's proposal for Saudi Arabia and Kuwait to recycle the dollars as low-interest-rate loans to the IMF. In addition, the work and testimony of John Perkins suggest that ensnaring governments by enormous debt in the global south was intentional. He testified that his job was

> to encourage world leaders to become part of a vast network that promotes U.S. commercial interests. In the end, those leaders become ensnared in a web of debt that ensures their loyalty. We can draw on them whenever we desire—to satisfy our political, economic, or military needs. In turn, they bolster their political positions by bringing industrial parks, power plants, and airports to their people. The owners of U.S. engineering/construction companies become fabulously wealthy. (2004, xi)

Research by the former chief economist of McKinsey corroborates Perkin's exposé and is worth quoting at length given the scope of the problem:

> Hundreds of billions of Third World loans were devoted to nonproductive projects and corruption. Many of these debt-financed projects also had harmful long-term

> consequences. In some cases, the chicanery took place on a purely local level. But what is most striking are the recurrent global patterns—overpricing, rigged bids, endless delays, loans to front companies with close ties to the government, investments in dubious technologies, 'public' projects undertaken for private motives, and private debts assumed by the state. Over and over again, we see the handiwork of the very same international banks, contractors, equipment vendors and export credit agencies, which grew fat while the countries grew poorer. These were not ideological errors—regimes of different ideological hues proved equally vulnerable. Nor were they due to random policy mistakes or purely indigenous corruption. A sophisticated *transnational system* of influential institutions contrived to produce similar mistakes over and over again, in every region of the world. Corruption has always existed, but without this global system, the abuses simply could not have been generalized on such a massive scale. (Henry 2003, xxviii)

Whether one believes that the deliberate inflation of the oil price and the rush to lend to the Third World were, in part, designed to trap the developing world in perpetual debt, there can be little doubt that private benefits accrued to financiers, construction and engineering firms, as well as the owning families of the oil and arms industry. Moreover, although the intentions can be debated, the effects of these actions are unmistakable. According to Stavrianos, debt in the developing world increased from '$19 billion in 1960, to $64 billion in 1970 and to $376 billion in 1979'—an increase of 1,879 percent over the period (1981, 448). At first these loans were enticing to governments and private enterprise in the Third World because they carried little interest and could be used to purchase arms, oil, industrial goods and construction contracts. However, the loans carried variable interest and in 1981, Paul Volcker, then chairman of the Federal Reserve, increased interest rates to unprecedented and unimaginable levels. Ostensibly this was done to stamp out inflation by sabotaging the US money supply, making it more expensive to take out loans and drawing money into the US Treasury through the foreign purchase of securities. But the over 400 percent increase in the price of oil was the overwhelming cause of high price inflation, and it seems indisputable that forces inside and outside the government were behind encouraging the increase of oil prices. In this light, it was obvious how to stop inflation: quit generating instability in the Middle East by militarizing the region, quit stalling on the Palestinian question, quit promoting coups and last, quit tolerating OPEC price increases. But halting these processes was never Washington's intention; it seems its intention was to see its oil companies prosper from price increases under the cover of OPEC while its major banks in the United States and the City of London searched for profitable avenues to make loans—the main source of their income.

But although this seems fairly clear from the available evidence, to the best of my knowledge, no one has looked closely at the data to see whether or not the unprecedented increase in interest rates actually did severely curtail the growth in the US money supply and a drop in the Consumer Price Index (CPI) or whether the massive increase in interest rates was enacted to render countries of the Third World permanently indebted to Washington DC and City of London bankers. Figure 5.3 shows the yearly averages for the growth of M2 from 1959–1985 to capture if there are any radical changes in the 1970s.

The trend is clear: a constantly expanding money supply at what appears to be a steady rate or a percent increase average from the beginning to the end of our period of about 9 percent. Figure 5.3 also superimposes the percentage change in the money supply as yearly averages and from year to year. What is perhaps most interesting to note is that there never appears to be a giant spike in the US money supply while the giant percent increase happened in 1967, not in in the 1970s. What is perhaps most remarkable is the fact that the percentage increases were in decline years before Volcker shocked the world with exceptional interest rates. Thus, we see no major spike in M2 and percentage increases in supply were in decline. In fact, an uptick in the percentage increases of the money supply happens exactly at the time of the Volcker shocks.

Figure 5.4 is perhaps even more suggestive. The conventional wisdom has it that high interest rates should lead to a drop in the money supply because loans become more expensive as interest rates increase and people shy away from expensive debt. Figure 5.4 appears to demonstrate that an increase in interest rates—even giant increases in the cost of money—have little effect on the general trend of M2 growth. Now, figure 5.5 does seem to suggest that

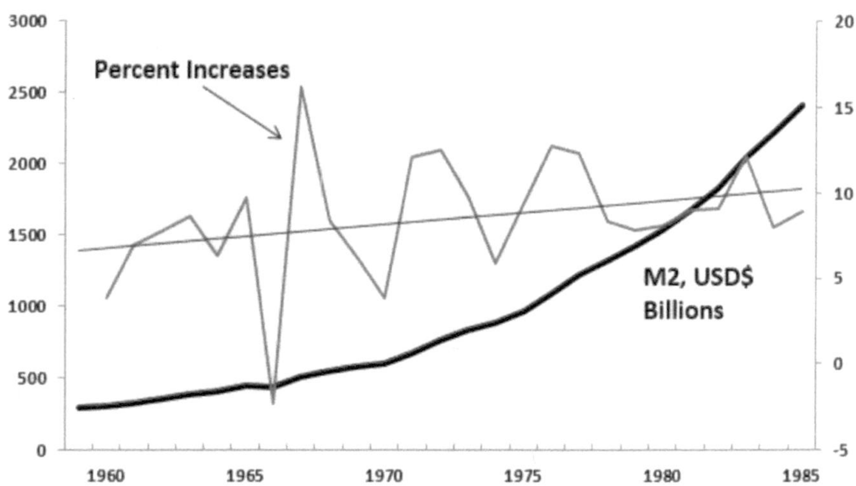

Figure 5.3. United States M2 Money Stock, Yearly Average and Yearly Percent Increases, 1959–1985. Source: Federal Reserve Bank of St. Louis. Trend line is for M2.

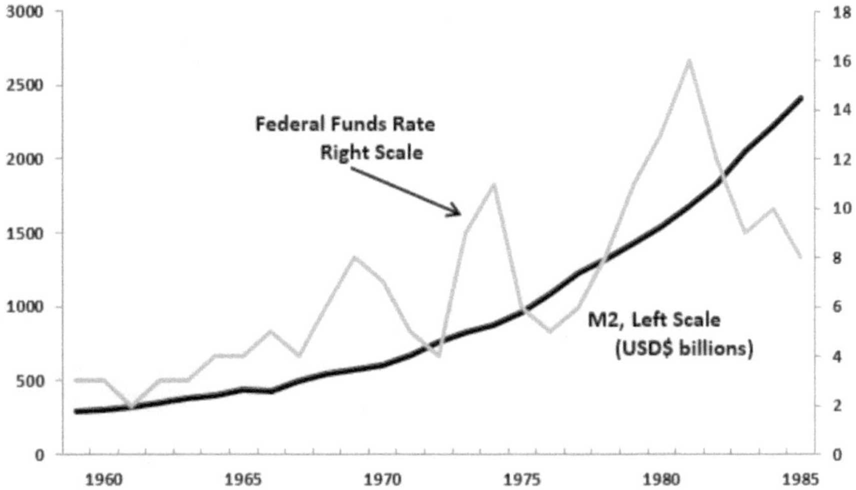

Figure 5.4. United States M2 and Federal Funds Rate, Yearly Averages, 1959–1985. Source: Federal Reserve Bank of St. Louis

there is a relationship between increasing the money supply and a general increase in the CPI. But the real question for the theory to hold—that is, for the thesis that Volcker increased interest rates to dampen inflation to be true—we would expect to see CPI either declining or staying steady as the rate of interest increases.

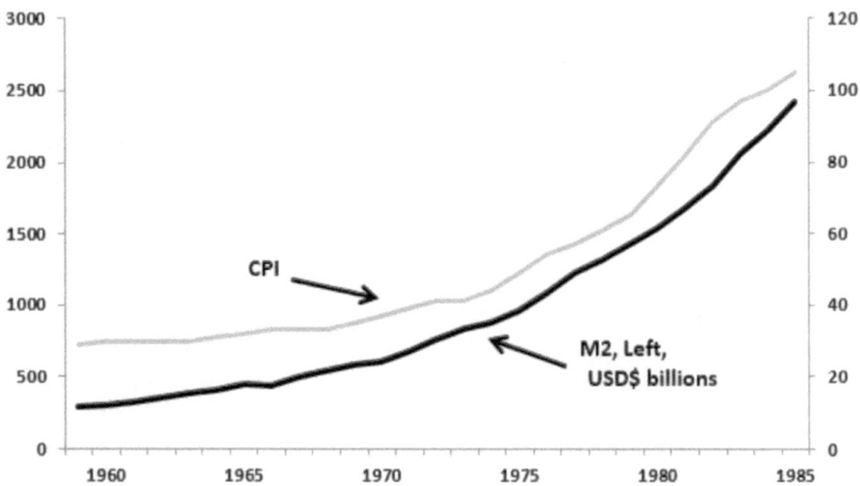

Figure 5.5. United States M2 and Consumer Price Index, All Urban Yearly Averages, 1959–1985. Source: Federal Reserve Bank of St. Louis

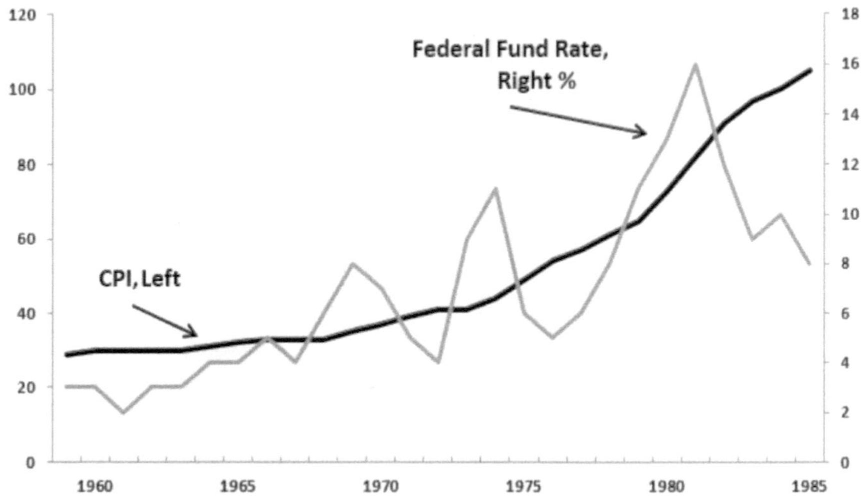

Figure 5.6. United States Consumer Price Index and Federal Funds Rate, 1959–1985. Source: Federal Reserve Bank of St. Louis

But the problem is that we do not see this at all from the data in figure 5.6. In fact, what is worrisome is that it looks like increasing interest rates may actually boost CPI if anything. This needs further study given that the evidence flails in the face of conventional wisdom. But a working hypothesis is that this should not be too surprising. We know M2 grew despite increases in interest rates, and for businesses, interest is a cost. Even when companies borrow at interest, they eventually must pass the cost of borrowing money onto consumers when they sell their goods or services. In this sense, and since the main way new money enters an economy is through interest-bearing loans from commercial banks, we should not be surprised to find that CPI increases over the long-term with positive interest rates. Figure 5.7 charts a closer look at the relationship between the yearly percent changes in M2 and the federal funds rate.

It is clear that there is some relationship between the two time series and the figure does suggest that (particularly after 1966) increases in the federal funds rate likely slow the rate of growth in the money supply. But notice the trendline for M2 in the chart. Although higher interest rates may affect the percentage change in the money supply to some extent, the money supply continues to increase. As stated earlier, there is an overall percent increase of about 9 percent over the period. So, although the federal funds rate may appear to have an impact on the *rate* of growth of M2, it does not change the fact that M2 continues to grow over the long term. A final point is to compare the massive drop in the rate of M2 growth in 1966 (–2 percent) when the federal fund rate was 4 percent with the period from 1979–1985, when interest

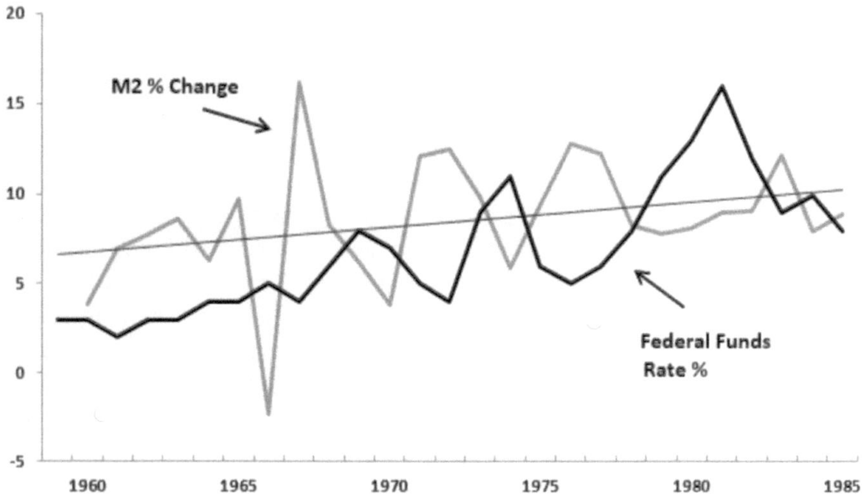

Figure 5.7. M2 Percent Change and Federal Funds Rate. Yearly Averages, 1959–1985. Source: Federal Reserve Bank of St. Louis

rates were hovering on yearly averages between 9 and 16 percent. During this period, the rate of growth in M2 was not negative as in 1966, but hovered between 8 and 12 percent increases per year. What this data seem to suggest is that, other than in rhetoric, increasing the federal funds rate had little impact on the growth of M2 or CPI. And this suggests another motive for increasing the interest rate to extraordinary levels. Either the lords of finance at the US Treasury, Central Bank and Wall Street somehow ignored oil price data and their correlation with inflation on the one hand, and on the other, could not foresee that unprecedented increases in the interest rate would massively push up the cost of loans everywhere contracts for debt in dollars were signed, or we must at least consider the more plausible thesis that the interest rate was used as a weapon of foreign policy. It was used as a weapon against the populations of the Third World and the working classes of the First World, not only to redistribute money to the owners of banks, but more importantly, to ensnare whole populations in unrepayable debt; to make the workers of the world and the countries of the global south 'revolvers'—subjects who service their debts but who can never repay them. By 1982, Mexico was unable to service its debts, and eventually the IMF was given the task of arranging new loans to restructure the debt, thereby creating a permanent and perpetual cycle of debt servicing. In return for these new loans, the IMF was also delegated with the power of commanding harsh economic reforms to ensure that public authorities could service its debts to private creditors (George 1988). In current scholarship, these reforms are called 'neoliberal' because they generally advocate policies or actions that promote greater market discipline and dependence for governments and their populations.[12] The policy suite of ten principles was dubbed the Washington Consensus by John Williamson in 1990 to capture the standard package of reforms applied to countries with the support of the US Treasury, the World Bank and the IMF. The suite of policies included the privatization of state assets, fiscal discipline (but no cuts to military budgets), the liberalization of trade, greater deregulation of business, positive interest rates and increasing taxes by broadening the tax base, to name some of the most familiar. The impact that these reforms had on the peoples of the developing world, and particularly its most vulnerable, has been studied by many scholars, but Abouharb and Cingranelli closely studied the impacts on human rights. The following excerpt provides a brief glimpse of what they found:

> Our findings confirm that the implementation of structural adjustment agreements leads to less respect for most but not all human rights we examined. More specifically, we show that governments undergoing structural adjustment for the longest periods of time have murdered, tortured, politically imprisoned, and disappeared more of their citizens. In addition, the execution of structural adjustment programs has caused governments to reduce their levels of respect for economic and social rights, created higher levels of civil conflict, and more abuse of internationally recognized worker rights. Our main argument linking structural adjustment policies to worsened human rights protection is that the policy changes implicitly or explicitly required in most structural adjustment agreements have hurt the poorest off in developing societies the most. (2007, 4)

Today there are 129 developing countries answerable to the World Bank's Debt Reporting System with a total external stock of debt now at US$4 trillion, up from US$1.9 trillion in 1995 despite some cancellation of debt through the Heavily Indebted Poor Country initiative of the 2000s. Brazil, Russia, India and China account for 40 percent of all external debt. The yearly interest charge has risen from US$85 billion in 1995 to US$155 billion in 2010. If we include principal repayments, the developing world collectively paid US$582 billion to their creditors in 2010, up from US$205 billion in 1995 (World Bank 2012, 40). It is also important to appreciate how increasing prices for oil will likely affect the global political economy in the future. At the time of this writing (January 2015), the price for a barrel of oil has dropped precipitously from over a hundred dollars in mid-2014 to just under US$48 dollars a barrel. A number of reasons are given for plummeting prices, from the economic slowdown in China to the ramping up of nonconventional sources of oil in the United States over the past decade (Porter 2015). However, some have suggested the Obama administration put pressure on Saudi Arabia to flood the market so that prices dropped enough to harm the economies of Russia and Iran, which are heavily dependent on oil revenues for state expenditure.[13] Others suggest that Saudi Arabia had its own motivations attached to maintaining its market share.[14] Still, as a nonrenewable resource, there can be little doubt that prices will eventually escalate again, and this means significant global redistribution from oil consumers to the owners of global oil (Akins 1972 was perhaps the first to warn of wealth redistribution). As the former director of the Central Intelligence Agency testified to the US House of Representatives: "[I]f the [energy] reserves of the Caspian Basin are added to those of the Persian Gulf, close to three-fourths of the world's oil, over the long term, will come from this region . . . as we move into the 21st century we are headed toward a massive transfer of the world's resources—hundreds of billions ranging toward trillions of dollars—into this volatile region [Caspian Basin and Persian Gulf]. Those funds will support much governmental and private activity that is not in the U.S. interest, to put it mildly."[15]

Thus, there is a looming question over what will happen when oil and gas are monetized in the future. How will this money be spent by the receiving countries? To be sure, a little over 90 percent of world oil resources are in the hands of state-controlled firms, leaving a little less than 10 percent of global oil resources in the hands of multinational firms.[16] Moreover, will importing countries be able to afford it, and what are the consequences of permanently higher prices for oil on the global economy? Hamilton (1983; 2009), for example, has already demonstrated that oil price spikes coincide with the major recessions of the twentieth century. Though this cannot be known with any certainty, a look at how the windfall of petrodollars was spent in the 2000s, as oil prices skyrocketed during the ongoing War on Terror may provide some clues. Although the data are patchy, the Federal Reserve of New York did track spending in 2006 and found that just under half the increase going to OPEC that year—US$475 billion—was used to finance imports and the other half—US$485 billion—used to purchase foreign assets or securities (Higgins et al. 2006, 2). As the price for a barrel of oil mounted throughout the 2000s, breaking the US$100 mark for the first time, a massive transfer of dollars flooded OPEC countries. From 2002 to 2013, OPEC countries spent US$40.8 billion dollars on arms transfers (constant US dollars). Figure 5.8 shows a time series of the largest purchasers by year.

The largest purchasers over the period were the United Arab Emirates at US$13.7 billion, Saudi Arabia at US$8 billion and Algeria at US$7.4 billion, or 71 percent of total spending by OPEC. The largest suppliers were the United States, Russia, Germany, France, the United Kingdom and China. Thus there is little doubt that there is a relationship between oil revenues gained by OPEC and the recycling of some of those proceeds in arms transfers. The question is who benefits from these transfers and what the amass-

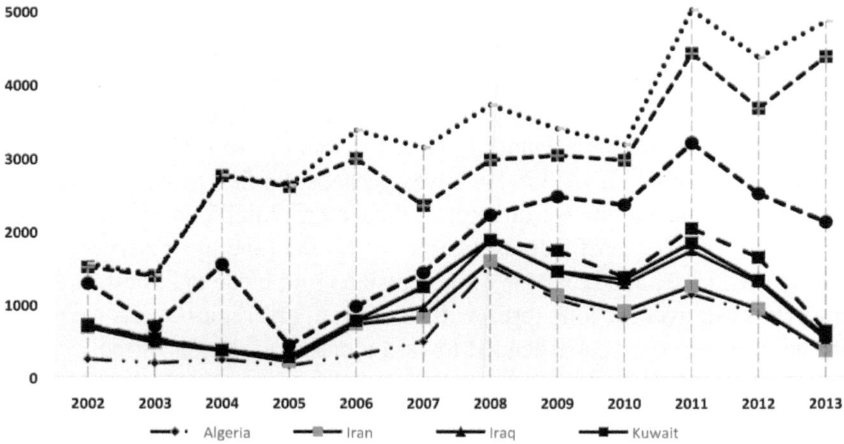

Figure 5.8. Leading Spenders on Arms Transfers, OPEC 2002–2013, US$ million, constant 1990. Source: SIRI Import/Export Database, http://armstrade.spiri.org/armstrade/page/values.php

ment of arms has on the prospects for future conflicts in the Middle East (recognizing, of course, that not all OPEC countries are from the region).

According to the two scholars who have done the most to advance our knowledge in this area, the main beneficiaries of these transfers, other than the regimes themselves, are the arms companies on the one hand and the oil companies on the other. Without conflict or the threat of conflict, the arms industry could not justify the massive public budgets that sustain research and development, as well as the manufacture of military technology. Yet, although the greater militarization of the world might not be the primary *cause* of conflicts and war, there is little doubt that having weapons and other tools of war contributes to and facilitates the likelihood of conflict and bellicosity (Bas and Coe 2012; Beker 1982; Hartung 1995; Kinsella 1998). Moreover, the leading seller of arms, the United States, has been transferring arms to human rights violators and dictators for decades (Blum 2004; Chomsky and Herman 1979). In testimony before the Subcommittee on International Operations and Human Rights, House International Relations Committee, one of the world's leading investigators noted that "from Iran to Indonesia, and from Central America to the Congo, our nation's role as the world's leading arms merchant has done far more harm than good. Using arms sales as a way to win friends and intimidate adversaries has not only fostered serious human rights abuses in the recipient nations; it has also undermined U.S. interests by spreading instability and fueling conflict."[17]

However, to the dominant owners of arms producers, the globalization of insecurity and arms has been necessary to maintain their profitability vis-á-vis their civilian competitors in the wider corporate universe. Arms manufactures are beholden to the same logic of differential accumulation as Apple or Google. Their level of capitalization relative to other firms is largely contingent on their rate of profitability. If armament firms consistently underperform the wider market, then their stock will undoubtedly take a battering as investors sell their shares in the hopes of making higher returns elsewhere. Thus, the executives and wage-dependent workers of arms companies not only have an interest in foreign sales (where they make better returns), but also in obtaining even more valuable government procurement contracts and, by extension, the *perpetual militarization* of world order. Peace is bad for business. As it turns out, and perhaps not surprisingly to those attuned to how conflicts in the Middle East and elsewhere drive share prices, the Bush and Obama administrations' War on Terror has been a bonanza for the owners of arms-producing and military services companies. Figure 5.9 shows the return on investment over the War on Terror period for the S&P 500, with an average rate of return a mere 7 percent.

Thus, had you invested US$100 in the overall index throughout this period, you would have made US$7 dollars at the end of the period—not a great return over fourteen years. Figure 5.10 plots the yearly share price of the five largest publically listed arms manufacturers in the United States ranked by the Stockholm International Peace Research Institute (SIPRI) according to sales (2013). The uptrend is clearly visible from the start of the War on Terror, and table 5.1 provides data on the overall returns from the start of the War on Terror to 2014.

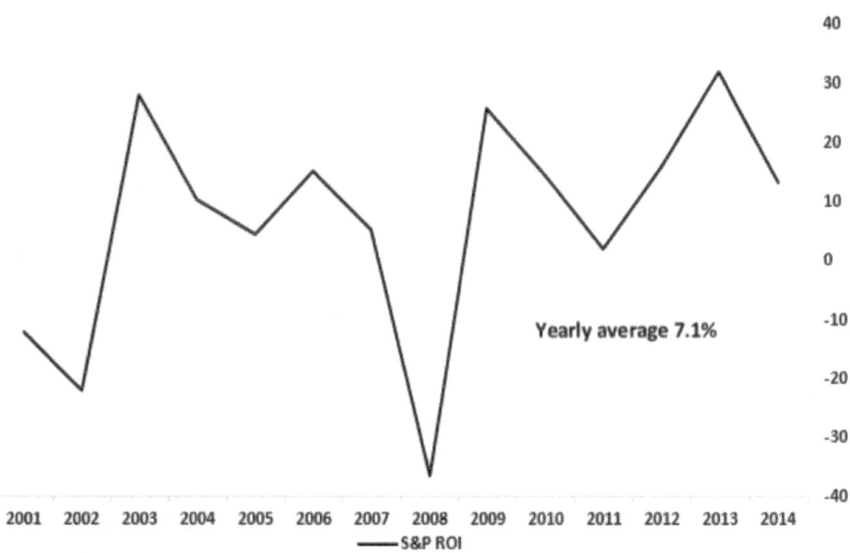

Figure 5.9. S&P 500 Returns on Investment, Yearly, 2001–2004

Table 5.1.

Returns on the Top 5 US Arms-Producing Companies 2001–2015	
Lockheed Martin	318%
Boeing	285%
Raytheon	192%
Northrop Grumman	188%
General Dynamics	189%
Average Return	234%

Compared with the return on investment in the S&P 500 index, the numbers are unmistakably stark. To take our previous example, if you would have invested US$100 dollars in an index of just these five companies, at the end of the period you would have earned US$234 dollars. But think about what this means for dominant owners who are far more financially invested than the average citizen in the United States and you can plainly see the class power dynamics at work. Eighty percent of the population owns only 5 percent of all outstanding financial wealth, whereas the top 1 percent own 42 percent and the next 19 percent of the population owns 55 percent of financial wealth (Wolff 2010). At the time of this writing, the former executive chairman of Lockheed Martin, Richard J. Stevens, was a direct holder of 208,086 shares in the company.[18] Suppose that he owned this many shares at the start of the Bush administration's War on Terror. Initially, Stevens's shares would have been worth about US$9 million dollars if they were cashed in. At the time of this writing, they are worth US$38 million—not a bad return over fourteen years of war. What also must be contemplated is the fact that the share prices appreciated as the war was prolonged. That is to say, Stevens and the other dominant owners of Lockheed Martin accumulated more money the longer the war dragged on. This is in no way to suggest that Lockheed Martin's executives prolonged the war, merely to state the factual evidence that the price of company shares appreciated remarkably throughout the war and likely would have not risen to such levels without the longest war in US history.

We do not have the space to show how other members of the 1 percent benefited from the armament firms, but we can estimate that with returns like those shown in table 5.1, dominant owners made out handsomely. Such returns make the phrase 'getting away with murder' rather trite and reminiscent of Marx's citation of T. J. Dunning, the secretary to the London Consolidated Society of Bookbinders: "Capital eschews no profit, or very small profit, just as Nature was formerly said to abhor a vacuum. With adequate profit, capital is very bold. A certain 10% will ensure its employment anywhere; 20% certain will produce eagerness; 50% positive audacity; 100% will make it ready to trample on all human laws; 300%, and there is not a crime at which it

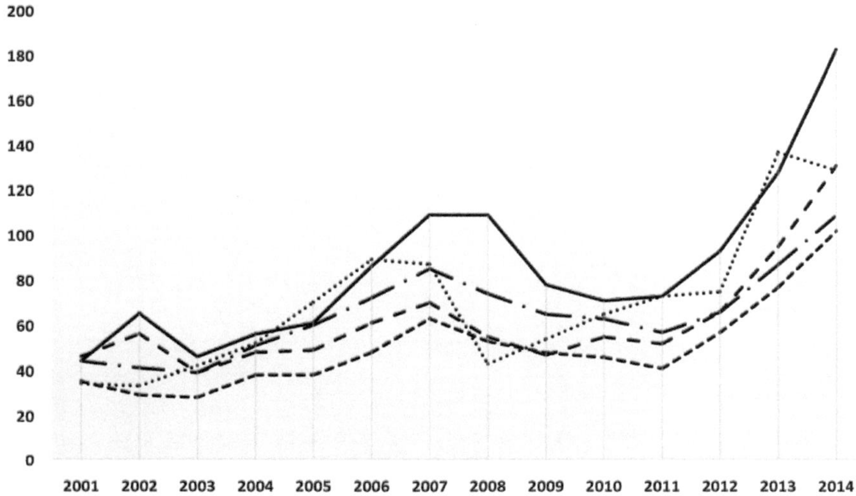

Figure 5.10. Share Price of Top 5 US Arms Firms, 2001–2004. Source: Google Finance

will scruple, nor a risk it will not run, even to the chances of its owner being hanged. If turbulence and strife will bring a profit, it will freely encourage both" (cited in Marx 1976, 926, fn 15).

Marx forgot to add that your chances of being hung are slim to none if you can effectively purchase the lawmakers of the country and stand by their side to promote 'security'. There is little doubt that the War on Terror caused much 'turbulence and strife' for many and great profit for the few. This tendency is also mirrored in the oil companies. When the official reasons for the invasion and occupation of Iraq were proved to be unambiguously false (that Saddam Hussein possessed weapons of mass destruction), critical journalists and scholars focused on how US and British corporations might get contracts to control the supply of Iraqi oil. They should have paid much more attention to the share price of the leading US oil companies as insecurity in the Middle East pushed up oil prices and corporate earnings into the stratosphere. Figure 5.11 shows the yearly changes in the share price of ExxonMobil and Chevron, the two largest companies by market capitalization from the United States.

With a dip during the global financial crisis recognized, the trend is clearly upward with investors gaining a return of 135 and 150 percent, respectively—that is, if they bought in 2001 and held their shares until January 2015. The chart also plots total capitalization for the oil and gas industry, which also increased during the War on Terror by 182 percent. The chairman, president and CEO of ExxonMobil is Rex Tillerson. Tillerson is considered a major holder of the company and owns 2,289,139 shares. If he owned that many

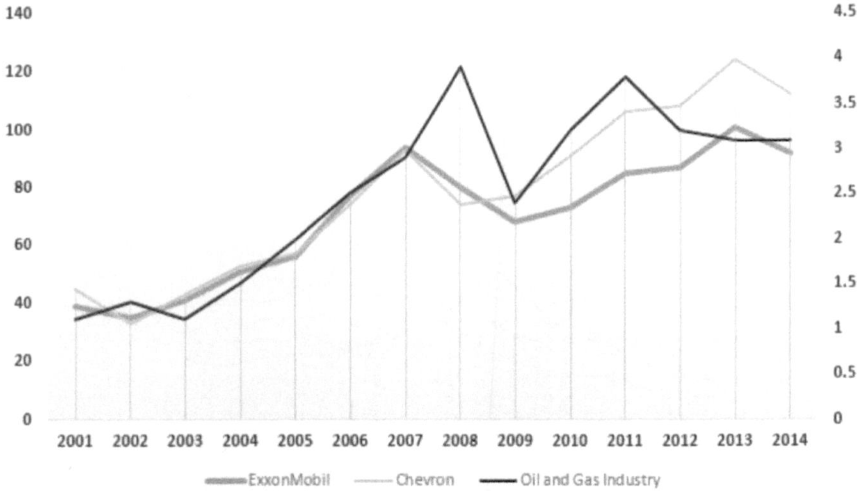

Figure 5.11. Increase in Share Prices for ExxonMobil and Chevron (left, US$) and Total Yearly Capitilazation of the Oil and Gas Industry (right, US$ trillions), 2001–2014. Source: Google Finance (yearly price shares), FT Global 500 (total capitalization)

shares in 2001, they would have been worth US$90 million dollars. After fourteen years of war, Tillerson's shares are now worth US$212 million. This relationship between conflict and rising capitalization for the oil and gas and armament firms could perhaps be considered an historical anomaly. Moreover, the ongoing conflict in the Middle East spurred largely by Anglo-American–made weapons and decades of strategic foreign oil policy in the region is likely not the sole cause of the increase in the price of oil, which contributed to the windfall profits of the oil and gas firms. But the connection between a belligerent foreign policy in the region and the differential profiteering by the owners of oil and gas companies and armament firms cannot be so easily dismissed as an historical aberration. Using data from the *Fortune 500*, Nitzan and Bichler (2004, chapter 5; 2014) have demonstrated that, with only one exception, every conflict in the Middle East has taken place when the leading oil and gas firms *de*accumulated relative to their corporate peers. Conflict and war in the Middle East were then followed by an upturn in the differential accumulation of the oil companies as returns on equity increased. Moreover, even if this observable relationship between conflict and increasing differential capitalization can be interpreted as somehow an historical accident (which is highly doubtful), it would seem strange that leading government officials and oil company executives do not know that conflict in oil regions generates higher prices for oil and therefore higher profits and share prices. And for the armament firms, the relationship is obvious: they require government largess and war or the threat of ongoing conflict in order to accumulate differentially, not to mention justify their continued existence. Threat inflation is a natural corollary of the vested interests in conflict and 'security'. Many of our problems are solvable, but what if those in power have no interest in solving them. What if all this global harm actually benefits a small coterie of people? But there are other beneficiaries of the War on Terror than the dominant owners of armament and oil firms: foreign and domestic banks and investors in US government securities. Since the barbarous attacks of 9/11, the federal debt of the Bush and Obama administrations pushed the national debt up by 197 percent from roughly US$6 to 17.8 trillion at the end of 2014.[19] As a proportion of this increase, military spending accounts for US$7.7 trillion during the same period, or about 43 percent of the increase in federal debt. Global turmoil also tends to promote the purchase of US securities as a safe haven for investment, drawing dollars back into the economy and boosting sales in US Treasury notes, bills and bonds, not to mention corporate bonds, securities and real estate.[20] But although all these purchases have impacts on the value of the dollar, and thus interest rates, what ultimately matters from the point of view of the capital-as-power perspective is how these deficits are capitalized for profit by investors and how this act capitalizes the power of the state to pursue its belligerent foreign policies in the Middle East and maintain its energy-intensive economy at home. Thus purchasing US federal securities is also a vote of confidence in the foreign and domestic policies of US elites given that research shows that the majority of US citizens have virtually zero say over policy (Di Muzio 2007; Gilens and Page 2014). Moreover, there are only two main ways in which the money supply can be increased in the United States. The first is by commercial banks making loans to its customers premised upon the creditworthiness of the borrower. In the US banking system, loans create deposits; deposits or reserves are never the basis on which money is created (Sheard 2013). This is what Ingham (2004) calls 'capitalist credit money', and the only limit placed on its creation is the ability of borrowers to repay the loan with interest. We also have to remember that there is always more debt in the economy than there is money to repay because interest is never created when loans are made, only the principal (Rowbotham 1998).

This system of money creation naturally produces inequality since the more wealth you have, the easier it is to borrow on better terms relative to those with no assets or those who make smaller incomes. It also tends to lead to massive asset bubbles as credit is dumped into real estate or the stock markets of the world rather than creating new productive capacity and jobs (Wolf 2014). Whereas billionaires may be able to take massive loans in the millions of dollars, an individual making a medium income, even in a rich country, will virtually never have access to that kind of credit. But aside from sustaining and amplifying inequality, many international commercial banks have also profited from the War on Terror and global climate change by helping to finance arms companies, oil and gas firms and other firms related to the war and reconstruction efforts. Hundreds of billions of dollars taken in loans throughout the War on Terror have left the oil industry with a growing debt owed to commercial banks and financial markets to the tune of US$2 trillion (Morales 2014). Moreover, commercial banks and other financial market operators also have positions in a trillion-dollar energy derivatives market. Many fear that if oil prices plunge too low, this debt will be increasingly difficult to repay and the derivatives market in energy may implode because of traders betting the wrong way on oil prices. Though it is difficult to tell with any certainty the amount of loans commercial banks extend to the arms industry, there is little doubt that it is significant and a chief way of increasing the money supply. For instance, a civil society movement to ban cluster munitions found that a consortium of 151 banks had invested US$27 billion dollars in cluster munitions manufacturers from 2011 to 2014.[21] Moreover, commercial banks often own considerable shares in arms manufacturers. For example, the Bank of New York Mellon owned 19.5 million shares in the top four US arms manufacturers valued at US$2.4 billion in January 2015. In the UK, a report found that 'four of the five major UK banks hold shares in every single one of the UK's top arms companies' with the fifth having a stake in six of the ten firms (War on Want 2008, 10). Nitzan and Bichler (2002) have explored the notion of an oil–arms nexus, but a more accurate focus might be tracing a bank–oil–arms nexus, with the Anglo-American central banks and their treasury departments at the epicenter of this network. The second main way that new money enters the economy is by governments running deficits and going into debt. I will focus here on the federal government of the United States for the simple fact that its budget is decisive for the national economy (as we saw in chapter 4, state debt is comparatively small). US treasuries largely consist of bills, notes and bonds, with the primary difference being their date of maturity. What few realize is that these transactions—the purchase of treasuries by the public—actually destroys rather than creates money. Taxation also destroys money; it does not give the federal government money to spend as is commonly believed. Its main function in the reserve system is to create and maintain a demand for the government's currency and destroy excess reserves (Bell 2000, 614). The reason for this is how the balance sheets of reserve accounting work in a fiat money system—that is, a money system backed only by sovereign enforceability. As the important work of Bell (2000) points out, '[T]he government finances all of its spending *through the direct creation of (high-powered) money*' by drawing on its account at the Federal Reserve. This spending immediately becomes part of the money supply, M1 if it is deposited into checking accounts and M2 if it is deposited into a savings account (Bell 2000, 615, my emphasis). The government does not have to sell treasuries, but because the way the system was designed, if it wants to maintain a positive interest rate, the government is forced to sell securities to the financial industry and public. A positive interest rate provides capitalists with a benchmark interest rate, necessary for providing a yardstick for differential accumulation. This is what is traditionally known as the 'normal rate of return' and is essentially risk-free profits handed over to the owners of commercial banks. Banks that hold excess reserves—that is, banks that have more reserves than they need to hold by law—can use these reserves to purchase treasuries rather than hold non-interest–bearing money. It also allows other institutions such as insurance firms a relatively safe place to park their money and make a return while doing so.

For example, Warren Buffet's main business, Berkshire Hathaway, takes in money from people who pay car insurance premiums and then he invests a portion of those premiums in government treasuries among other investment vehicles. If this safe haven did not exist, Buffet would have to put all of his resources in more risky (but potentially more rewarding) financial instruments. Once we realize how the reserve accounts actually work in practice, it becomes plain to see that 'fiscal policy has more, and monetary policy less, to do with the money supply than is usually recognized' (Bell 2000, 216). To put it in simple terms, it matters very much what the government spends money on since this is one of the chief ways in which the money supply can be extended. As we have seen, a major priority of the US government since World War II has been on war spending, with the Pentagon the single largest consumer of oil on the planet. Unknowingly, President Eisenhower had spelled out what this spending pattern has meant for the differential accumulation of dominant owners and the broader consequences for the American and global public:

> Every gun that is made, every warship launched, every rocket fired signifies, in the final sense, a theft from those who hunger and are not fed, those who are cold and are not clothed. This world in arms is not spending money alone. It is spending the sweat of its laborers, the genius of its scientists, the hopes of its children. The cost of one modern heavy bomber is this: a modern brick school in more than 30 cities. It is two electric power plants, each serving a town of 60,000 population. It is two fine, fully equipped hospitals. It is some fifty miles of concrete pavement. We pay for a single fighter plane with a half million bushels of wheat. We pay for a single destroyer with new homes that could have housed more than 8,000 people. This is, I repeat, the best way of life to be found on the road the world has been taking. This is not a way of life at all, in any true

sense. Under the cloud of threatening war, it is humanity hanging from a cross of iron . . . is there no other way the world may live?[22]

Knowing the priorities of the arms and oil and gas industry and the banks who either own shares in these companies and/or make loans to them, there may be 'no other way the world may live'. Given our empirical analysis thus far, it can hardly be doubted: they have little stake in changing courses. We have already argued that it is the dominant owners of banks, arms manufacturers and oil men (and they are mostly men) who benefit from war spending.

Figure 5.12 shows the historical relationship between the US federal debt and defense spending with the two series highly correlated. The story in figure 5.13 appears to be rather straightforward. With some exceptions (1969 and from 1998–2001), the US federal government started running consistent budget deficits, spending more money into the economy than it received in revenue. The relationship between the deficit and US defense spending is clearly inverse, with the major surpluses of the late 1990s corresponding with a historical drop, albeit minute, in military spending. As spending on defence increases, so, too, does the 'national' deficit. Massive defense spending is, of course, not the only cause of persistent deficits in the United States. For instance, much of the trillion-dollar deficits from 2009 to 2012 can be attributed to the federal government's bailouts of the banks and other companies largely owned by the top 20 percent of the US population. Still, given the relationship illustrated here and the fact that US defence spending accounts for 19 percent of the federal budget, curtailing military spending would make a significant

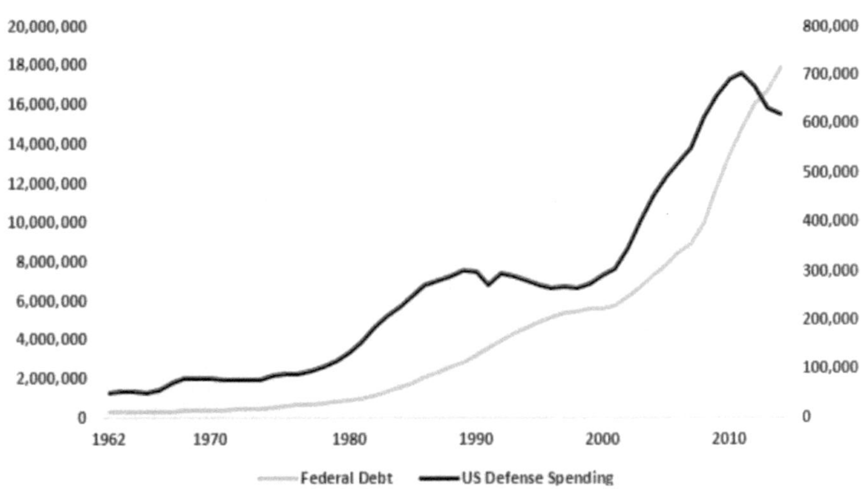

Figure 5.12. US Defense Spending and US Federal Debt, 1962–2014. Source: Office of Management and Budget, http://www.whitehouse.gov/amb/budget/Historicals

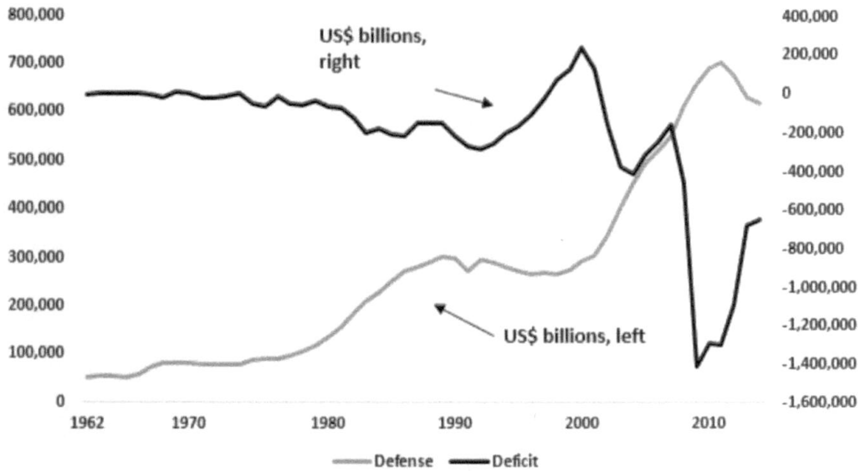

Figure 5.13. US Defense Spending and US Federal Budget Deficits, 1962–2014. Source: Office of Management and Budget, http://www.whitehouse.gov/amb/budget/Historicals

difference in balancing the books or the ability to finance more humane projects of benefit to all US citizens. Figure 5.14 also demonstrates that the historical relationship between the federal debt of the United States and the money supply and shows them to be positively correlated so that military spending, in part, becomes one of the chief ways of expanding the money supply.

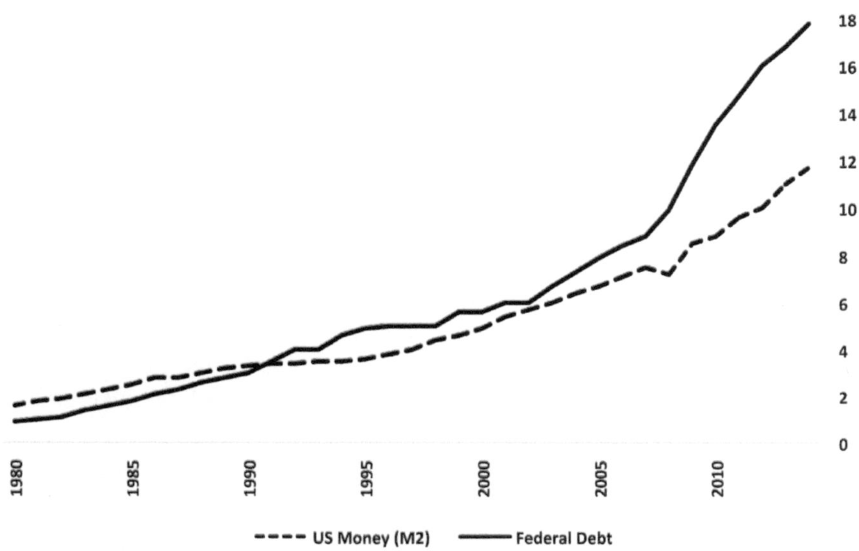

Figure 5.14. US Money Supply (M2) and US Federal Debt, US$ trillion. Source: Federal Reserve Bank of St. Louis and Office of Management and Budget

Table 5.2.

Ownership of US Treasuries Fiscal Year End, 2013[1] Debt Value: US$17.4 trillion	
Publically held	
SOMA and Intragovernmental holdings (incl. federal reserve banks)	US$7.2tr
Total Public	US$7.5tr
Privately Held	
Depository institutions	US$321bn
US savings bonds	US$179bn
Private (pension)	US$492bn
State and local governments (pension)	US$203bn
Insurance companies	US$264bn
Mutual funds	US$1.1tr
State and local governments	US$593bn
Foreign and international	US$5.7tr
Other investors	US$1.1tr
Total Private Held	US$10.1tr

[1] Data are from the US Treasury. 2013 is used rather than 2014 because this was the last year the information was available for all owners (http://www.treasurydirect.gov/govt/resources/faq/faq_publicdebt.htm#DebtOwner).

But what does the ownership distribution of US treasuries look like? Table 5.2 breaks down the ownership of US treasuries into aggregates.

Note that these aggregates hide a considerable amount of detail so that it is unclear exactly who owns the debt as individuals and families. This is part of the secrecy of ownership under carbon capitalism, which serves to mask the domination of the majority of the world's population by a tiny amount of dominant owners. Note that the majority of US treasuries are privately held. However, research on the distribution of ownership has been sparse in the scholarly literature until recently. The pioneering work of Sandy Hager (2013a) helped to illuminate our understanding of the distribution of this debt and its primary beneficiaries. Albeit with some caveats due to a lack of data in the 1970s, Hager found that ownership is heavily concentrated in the top 1 percent of households at 42 percent. Moreover, Hager's research also demonstrated that 'what the top one percent gives to the federal government in income taxes as a percentage of its income has, at least since the early 1980s, failed to keep pace with what it receives in federal interest payments' (2003, 174). Even when considering intragovernmental transfers, which are typically believed to help bottom income earners, Hager's evidence suggests the conclusion that 'the public debt has come to serve as an institution of power working in the interests of the top one per cent' (2003, 177). In other words, the massive debt incurred by the War on Terror for military spending not only helped to increase the money supply to enrich the owners of arms and oil companies, but also to redistribute income to the top 1 percent of bondholders. Perhaps the only development that could potentially challenge the current power dynamics of differential accumulation is the renewable energy industry. In the next section we will consider this prospect based on the available evidence.

CAPITALIZING FOSSIL FUELS AND THE PROSPECTS OF A RENEWABLE ENERGY FUTURE

Whether implicitly or explicitly, the majority of humanity is resting its hopes on a renewable energy future. Thanks in part to the advertising and marketing campaigns of energy firms, many believe that the world is getting greener and cleaner, developing toward an alternative energy future beyond fossil fuels and the quest for never-ending growth. The mainstream media have not been very helpful in critically scrutinizing this vision of the future. They emphasize the return of electric motor vehicles, hybrid electric/petrol cars like the Prius, more wind turbines, solar cells and biofuels, not to mention a surge of investment in alternative energy companies. Moreover, mounting oil prices throughout the era of the War on Terror and the subsequent global financial crisis of 2007–2008 eventually provoked many governments to support the nascent renewable energy industry. If all these developments occurred across the world in the 1970s and were combined with the support for efficient public transit systems, the promotion of the bicycle as a primary means of transport (where feasible), the support for more local food systems, penalties for corporations engaged in planned obsolescence or a ban on the practice altogether and urban planning focused on energy efficiency and the human scale rather than scaling cities and suburbs for automobiles and trucks, there is little doubt that global society would be on a much more sustainable path. But this did not happen. As we have seen earlier, the carbonization of social reproduction continued to globalize and intensify since the 1970s. The actions taken by the executives of oil companies, banks, automotive companies and real estate developers and the politicians they were able to influence served to incorporate more and more of the world's population into an unsustainable project of mounting mass consumerism founded on nonrenewable and environmentally ruinous fossil fuels. To put things in scientific perspective, consider the scale of the project of replacing nonrenewable fossil fuels with an alternative energy source such as direct sunlight: "In order to produce enough [solar] energy over the next 25 years to replace most of what is supplied by fossil fuels, the world would need to build 200 square meters of solar photovoltaic panels every second plus 100 square meters of solar thermal every second plus 24 3-megawatt wind turbines every hour nonstop for the next 25 years" (Assadourian 2010, 7).

As the informed reader will be able to tell, such a project is not presently underway despite the fact that leading policymakers are keenly aware of the scale of the problem. In addition, energy consumption is expected to increase by some 37 percent by 2040. However, without widespread efficiency gains in infrastructure, industry and vehicles, ExxonMobil estimates that energy demand will increase by some 140 percent by 2040 (2014, 7). Continuing to put things in perspective, consider that the IEA estimates that at least US$48 trillion will be needed to supply the world's growing thirst for energy by 2035. US$23 trillion of this massive sum will be used to extract, refine and transport fossil fuels (IEA 2014a, 1; 2014b, 11). The same report estimates that US$6 trillion needs to be invested in renewables, mostly for electricity generation. To get a sense of the scale of this level of investment, consider that US gross domestic product (GDP) is about 14 trillion dollars, whereas the market capitalization of most every publically listed company is roughly US$65 trillion. You can start to see that US$48 trillion is an astronomical sum.

In 2011, when the latest figures were available, the total production of global energy from renewable sources was 18.4 percent, or 13.3 percent if we leave out power generated from uranium, a fairly abundant source, but one that is ultimately nonrenewable, not to mention fraught with ecological dangers, from storing spent fuel rods to the potential for accidental meltdowns like in Chernobyl and Fukushima (IAE 2013, 6). Reaching about 20 percent renewable energy might sound impressive to some, but when compared to the 14.4 percent of total primary energy supply produced in 1973, it becomes clear that progress in renewable energy has travelled at the pace of a snail. One of the many reasons for this is the subsidies the fossil fuel industry receives compared to renewable energy technologies. In 2013 alone, the fossil fuel industry received US$550 billion in government subsidies, whereas the renewable energy industry received US$120 billion (IEA 2014b, 6). So far the IEA has reported that from 2007 to 2013, cumulative investment in renewable power capacity has been US$1.5 trillion and anticipates that this will barely increase from 2014 to 2020. Moreover, according to the forecasts made by ExxonMobil (2014), fossil fuels will continue to dominate energy demand in 2040 at 77 percent of total demand.

The powers of the fossil fuel industry, the lack of significant investment and the disproportionate subsidies to each industry are not the only reasons that a post-carbon era founded on renewable energy is being forestalled.

There appear to be two more highly significant reasons. The first is that, at present, there is significant evidence to suggest that current forms of energy-intensive social reproduction cannot be sustained with known sources of renewable energy. At best, renewable energy may make up less than 30 percent of the world's energy supply in the twenty-first century (Heinberg 2009; Smil 2011; Trainer 2007; Zehner 2012). The difficulties rolling out renewable energy have been discussed at length in the literature, so there is little need for me to repeat what has already been written. Therefore, I will only summarize some of the main factors that hamper the renewable energy industry. Some of the main problems include:

- Reliability, as some sources are intermittent (e.g., wind and tides)
- Scalability (e.g., wind turbines and solar cells)
- The conversion of fertile land to wind farms and/or biofuels (e.g., the loss of food crops)
- Negligible energy returned on the energy invested (e.g., some biofuels)
- Integration into pre-existing power infrastructure (e.g., electricity generated by wind)
- The inefficiency of battery storage (e.g., the loss of energy during conversion)
- High cost to consumers (e.g., the price point of photovoltaic solar cells)
- Components made with exhaustible and expensive rare earth elements (e.g., gallium and indium used in solar cells)
- Low winter insolation, dust and water vapor and clouds (e.g., photovoltaic solar cells)
- High capital investment (e.g., wind turbines, solar cells, research and development)

These are just some of the major problems with renewable forms of energy. However, even if some of these obstacles could be overcome or some miracle energy source was found that emitted zero carbon, the problem of overconsuming the planet's resources by the rich and affluent consumer classes would still be problematic on a planet with many finite or exhaustible resources.

But there is a second and perhaps more significant reason why renewable energy is experiencing relatively slow growth: the simple fact that it is not very profitable relative to the oil and gas industry. Recall that investors capitalize expected future profits. In this sense, 'market capitalization contains a forward-looking element, as share prices include a view on investors' expectations'.[23] What the ritual of capitalization suggests is that we should consider the market capitalization of energy firms as a leading indicator of what capitalists

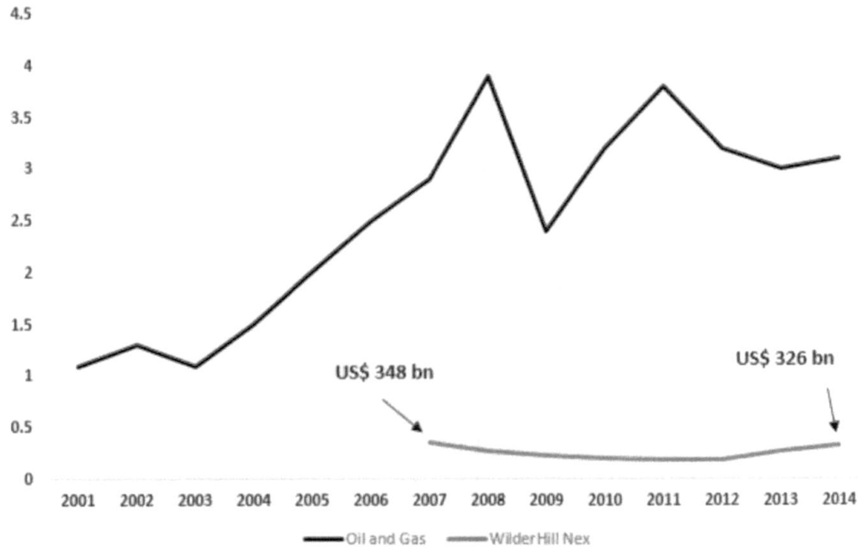

Figure 5.15. Oil and Gas Company Capitalization vs. WilderHill Nex, 2001, 2014. Source: FT Global 500 and WilderHill Nex

think will happen in the future. Figure 5.15 charts the capitalization of the leading oil and gas firms that are publically listed on international exchanges and compares this to the capitalization of companies that compose the WilderHill NEX.

The latter 'is a global index of 106 companies listed on 31 exchanges in 26 countries (excluding Hong Kong and Taiwan) whose innovative technologies and services focus on the generation and use of cleaner energy, conservation, efficiency and the advancement of renewable energy in general' (WilderHill NEX 2014). Put simply, the index should be a strong indicator of how well the renewable energy industry is performing as a sector of the global economy. The evidence in the chart tells a relatively straightforward story. At the beginning of the period, the industry had slightly over US$1 trillion in market capitalization. This escalated to a high of US$3.9 trillion by 2008 until market capitalization collapsed to US$2.4 trillion during the global financial crisis. After the crisis, capitalization almost recovered to its previous high and then started to taper off toward the end of the period. However, the capitalization of the leading oil and gas firms was still US$3.1 trillion in 2014. Now consider the capitalization registered by the WilderHill Index. At the start of our period (for when there is data), the value of all the firms gathered by the index was US$348 billion. After this high point, the capitalization of cleaner energy firms began to decline until a small uptick in 2014. Thus, we do not see any significant spike in capitalization that would suggest to us that investors envision a profitable renewable energy future. In fact, we almost see the exact opposite. Not only is the oil and gas industry's capitalization about ten times the value of the renewable energy sector (and much higher if we include the estimated market value of state-run oil and gas firms; see Di Muzio 2012), but the returns on investment are far greater. If the oil and gas capitalization charted here were an index, the return on investment would have been 181 percent, whereas the return on investment for the WilderHill NEX would be –6 percent over the period. It does not take a savvy investor to figure out which return is more desirable. Moreover, there is not one single firm represented in the top 500 companies by market capitalization on the *Financial Times Global 500*. Indeed, a report by Bloomberg New Energy Finance and the United Nations Environment Programme (UNEP) noted that "it is striking how poorly clean energy companies are represented among the world's largest stocks by market capitalization . . . the message appears to be that investors do not expect leading clean energy companies to grow at the same rate that the sector as a whole did between 2004 and 2009" (UNEP and Bloomberg New Energy Finance 2010, 23).

Nor are there encouraging signs in the immediate future. As an updated report on renewable energy investment from the IEA stated:

> Still, the capital-intensive nature of projects can make the risk/return profile of such assets challenging for investors. In 2013, global new investment in renewable power capacity was estimated over USD 250 billion, down slightly versus that in 2012, and lower than the near USD 280 billion registered in 2011. Despite overall higher global capacity additions (123 GW), declining unit investment costs in solar PV and onshore wind put downward pressure on investment levels. Over the medium term, annual investment in new renewable power capacity is seen averaging a somewhat lower level, at above USD 230 billion annually, in real terms, through 2020. (2014, 8)

Thus, one significant obstacle to the renewable energy industry is that its direct competitors in oil and gas are far more profitable. Though there will undoubtedly be ebbs and flows in the price of oil throughout the twenty-first century, there appears to be little doubt that, on average, prices for oil (and perhaps natural gas and coal later) will increase over time. According to the industry's authoritative *BP Statistical Review of Energy*, there are about 1.6 trillion barrels of proven oil reserves. Consider the temptation of selling these reserves. Even at an average price of US$100 for a barrel of oil, if all reserves were monetized today, you would have a figure of US$160 trillion. Even if we allowed for a meagre 10 percent profit margin on this sum, this would still mean that dominant owners of oil would split the spoils of US$16 trillion. To get some sense of scale, consider the fact that this sum is about equal to all the economic activity in the United States for one year. Of course, if the price shot well above one hundred dollars, as it is anticipated to do by many financial analysts, then the sums become astronomical. You can bet that the dominant owners of oil, not to mention gas and coal, have a financial interest in selling the last barrel of proven reserves. As previously stated, this will mean the complete destruction of the biosphere and a very different climate from what humanity has been accustomed to. In the next chapter we consider some of the dimensions of this post-carbon order.

NOTES

1. http://www.wired.com/2012/08/the-non-lethal-weapons-summit/.

2. Carola Hoyos, 'The New Seven Sisters: Oil and Gas Giants Dwarf Western Rivals', *Financial Times*, 12 March 2007.

3. All citations are from Carola Hoyos, 'The New Seven Sisters: Oil and Gas Giants Dwarf Western Rivals', *Financial Times*, 12 March 2007.

4. http://www.eia.gov/countries/regions-topics.cfm?fips=OPEC BUILD CHART FOR NEXT SECTION.

5. This is my calculation based on a Forbes report that Saudi Arabia produced 12.7 million barrels a day, or 4.6 billion barrels for all of 2013. The Energy Information Agency (EIA) reported that Saudi Arabia's net export revenues were US$274 billion in 2013. If we assume a very conservative production cost of US$10, then each barrel in net oil exports would garner US$49 in profit (http://www.forbes.com/pictures/mef45egdfi/1-saudi-aramco-33/) and for OPEC revenues (http://www.eia.gov/countries/regions-topics.cfm?fips=OPEC) and anticipated production costs (http://www.eia.gov/tools/faqs/faq.cfm?id=367&t=6).

6. http://blogs.ft.com/beyond-brics/2013/04/05/iea-charts-point-to-bric-led-future/.

7. The General Agreement on Tariffs and Trade was also created during this period, later institutionalized as the World Trade Organization in 1995.

8. These imbalances, however, turned out to be a structural feature of the global economy.

9. Sterling oil could be considered the minor exception due to the monetization of Anglo-Persian's oil assets and later the North Sea find.

10. http://www.opec.org/opec_web/en/data_graphs/330.htm.

11. Oliver Morgan and Faisal Islam, 'Saudi Dove in the Oil Slick', *The Observer*, 14 January 2001. http://www.guardian.co.uk/business/2001/jan/14/globalrecession.oilandpetrol. During an interview with Morgan and Islam, the former Saudi oil minister Sheikh Ahmed Zaki Yamani stated: "I am 100 per cent sure that the Americans were behind the increase in the price of oil. The oil companies were in real trouble at that time, they had borrowed a lot of money and they needed a high oil price to save them". Engdahl (2004, 130ff) believes that the policy to increase prices was already formulated by the Bilderberg Group at their Saltsjöbaden Conference in May

1973. He argues that documents from the conference demonstrate that a 400 percent increase was desired so that a recycling program could be instituted by the major US and UK banks operating in New York and London.

12. Di Muzio and Robbins (2016) have called this the debt–neoliberalism–restructuring nexus.

13. Larry Elliot, 'Stakes Are High as US Plays the Oil Card against Iran and Russia', *The Guardian*, 14 November 2014. Mark Mazzetti, Eric Schmitt and David D. Kirkpatrick, 'Saudi Oil Is Seen as Lever to Pry Russian Support From Syria's Assad', *New York Times*, 3 February 2015.

14. Angli Raval. 2015. 'The Big Drop:Riyadh's Oil Gamble', *Financial Times*. 9 March 2015.

15. Testimony of R. James Woolsey, US House of Representatives Committee on National Security, 12 February 1998 (http://www.loyola.edu/departments/academics/political-science/strategic-intelligence/intel/19980212woolsey.html).

16. Sheila McNulty, 'Politics of Oil Seen as Threat to Supplies', *Financial Times*, 9 May 2007.

17. http://www.worldpolicy.org/projects/arms/reports/testimony030701.htm.

18. http://finance.yahoo.com/q/mh?s=LMT+Major+Holders.

19. US Office of Management and Budget and Ownership of Federal Securities at the US Department of Treasury (http://www.whitehouse.gov/omb/budget/Historicals; http://search.treasury.gov/search?affiliate=treasury&commit=Search&query=ownership%20of%20federal%20securities).

20. Many have worried about the status of the US dollar throughout the War on Terror and whether or not it would require high interest rates to attract the dollar reserves of foreign countries. So far, this has not been the case given alternative investment opportunities. However, as the dollar depreciated in the mid-2000s, more investors bought gold, pushing up the price from US$376 an ounce in 2003 to US$1837 an ounce by mid-2011 (https://www.bullionvault.com/gold-price-chart.do).

21. http://www.stopclustermunitions.org/en-gb/media/news/2014/banks-behaving-badly-us$27bn-invested-in-cluster-munition-producers-while-use-of-banned-weapon-continues-in-ukraine-and-syria.aspx.

22. Dwight D. Eisenhower, 'The Chance for Peace', 16 April 1953 (http://www.edchange.org/multicultural/speeches/ike_chance_for_peace.html).

23. Anne-Britt Dullforce, 'FT Global 500, 2010', *Financial Times*, 28 May 2010.

Conclusion

The Post-Carbon Era and the General Crisis of Social Reproduction

> Perseus wore a magic cap that the monsters he hunted down might not see him. We draw the magic cap down over eyes and ears as a make-believe that there are no monsters![1]

It seems banal to state but is perhaps worth repeating that the world we inhabit today was neither constructed overnight nor constructed by the universal consent of humanity. As I hoped to have demonstrated in the previous chapters, carbon capitalism and a more extensive global petro-market civilization were forged by the powerful logic of differential accumulation as the uneven accumulation of money for dominant owners became wedded to fossil fuels, the martial and fiscal power of the state, and a monetary system that largely expands through privately capitalized debt. There is little doubt that things could have been otherwise or that the powerful confronted resistance and opposition in their quest to monetize energy and the natural world in the pursuit of private power.

But there is little practical use in theorizing how things might have been different if fossil fuels had never been commercially extracted or if global society pursued what might be called a logic of livelihood and ecological sustainability rather than the logic of differential capitalization and the augmentation of capitalist power. Given present trends, this concluding chapter argues that world order as it is presently constituted (though always in flux) is likely to undergo a general or civilizational crisis of social reproduction. This crisis will be experienced unevenly, just as the benefits of the present world order are also shared unevenly both within and between societies. What I mean by a general crisis of social reproduction is a multiscalar, multidimensional, and internationally interconnected series of events whereby current patterns of energy-intensive production, consumption, and reproduction can no longer be sustained, let alone expanded. This will herald what I call a second great transformation in civilizational order. If the first great transformation was the metamorphosis of agrarian societies of low growth into more urban, market-dependent societies experiencing compound economic and population growth and some form of democratic or social planning after World War II, then the depletion and greater cost of fossil fuels over the coming decades will signal the unevenly experienced creeping end of the global petro-market civilization in the next century.

So far we have only broached the topic of peak oil, or more broadly, peak fossil fuels and global climate change. Given the litany of recent works on both subjects, I have little desire to retrace the entirety of the literature or to engage in debates on precisely when peak oil will materialize or climate change will reach a tipping point beyond human control. But given the subject matter of this study, I would be remiss if this book did not conclude with some observations from the capital-as-power perspective on the likely consequences of peak oil and climate change as they relate to a civilizational crisis of social reproduction. I propose that these observations can be investigated along six dimensions that seem critical to the expanded reproduction of carbon capitalism and an uneven, yet worldwide, petro-market. I separate them out analytically to highlight each dimension, but they are all interconnected or linked in complex ways so that they cannot be separated out in practice. These dimensions are food and water, science, health and population, employment and mobility, climate change, global capitalization, and finally, democracy and civil peace. I explore each of these dimensions with a view to providing the reader with a few final words on carbon capitalism in the twenty-first century.

PEAK FOOD, PEAK WATER

It is often downplayed, but an important dimension of World War II for the Nazis and Imperial Japan, not to mention other belligerents, was the need to garner more land for the production of food calories to feed growing populations in the cities whose diets increasingly consisted of meat (Collingham 2012). This was certainly not the only historical war humans have fought for territory that could produce food energy and other useful natural resources, but it does draw our attention to the fact that less than a century ago, the search for arable land was a major cause of the world's second most ferocious total war. Millions of soldiers and civilians were gunned down, butchered, gassed and massacred in the pursuit of power, food and the carbon fuel required for the greater industrialization of imperial sociality. It may seem strange then, for a portion of privileged humanity that food energy is in abundance and that wars for territory and food have seemingly declined, though not disappeared.

To examine the questions of food, we must first recognize that 99 percent of global food consumption comes from arable land, with the remaining 1 percent from freshwater lakes and the world's oceans (Pimentel and Pimentel 2008). But the way in which a considerable portion of the globe's food is grown today is radically different from historical production methods. Currently, the world's capitalized food regime is highly dependent on oil and natural gas for its social reproduction, from farm, lake or sea to table. As we have suggested in chapter 3, arguably the first and most prominent form of capitalization resulted from the control of the land and its products through enclosure and the state's protection of private ownership. This trend continued during the age of carbon energy to a point where a significant portion of humanity is almost completely dependent on market transactions for their dietary requirements. Thus, it was not just the land that was enclosed by various measures, but more importantly, the ability to produce one of the most fundamental human requirements: nutritious food. Today, the top ten firms the *Financial Times* characterizes as 'food producers' have a total market capitalization of USD$677 billion dollars, making it the sixteenth-largest sector of the global economy out of thirty-seven sectors.

To be sure, there are many more food producers that are not registered on the *Financial Times Global 500*, but these firms would have considerably less power to shape and reshape the global terrain of food provisioning for social reproduction than the global giants who largely promote Westernized diets centered on meat, salt, sugar, fat and caffeine (Moss 2014). What many do not realize is that these Westernized diets are saturated in fossil fuels at every step of the supply chain. By one estimate, the modern food system absorbs about ten calories of fossil fuel energy for every calorie of food energy created (Green 1978; Heinberg 2003; Pfeiffer 2006).[2] Indeed, not only is oil necessary to run the industrial equipment and farm machinery used to produce modern diets, but the fertilizers, herbicides and pesticides made necessary by the corporate transition to industrial farming are all produced with oil and/or natural gas. In addition to this, most of the world's food travels thousands of miles and is dried, refrigerated, packaged and transported by a global fleet of thousands of diesel-burning trucks, trains and turbine-spinning refrigerated cargo jets—the least efficient method of food transport. Thus, an important dimension of carbon capitalism is the fact that many citizens are essentially eating fossil fuels as part of their social reproduction: "A whole generation of citizens thought that the carrying capacity of the earth was proportional to the amount of land under cultivation and that higher efficiencies in using the energy of the sun had arrived. This is a sad hoax, for industrial man no longer eats potatoes made from solar energy, now he eats potatoes partly made with oil" (Odum quoted in Smil 1994).

Given the centrality of oil and gas to modern food production, the global military presence of the United States, mobilized in part to ensure oil remains a tradable global commodity, along with wars and conflict generated by the control of oil stocks and flows, can be understood as an integral part of what Phil McMichael (2009) has called the 'corporate food regime'. Viewed in this way, it is simply not the case that the US-led wars in the Persian Gulf were fought to guarantee a world market for oil to ensure private automobility, high consumptive living and the profitability of oil and gas companies. Indeed, these actions also help safeguard an energy-intensive global food regime for the privileged soaked in fertilizers and pesticides derived from petroleum and natural gas.

Although some have praised this system of food provisioning for increasing caloric yields, and some attribute it to the rise in world population without widespread famine, the so-called Green Revolution and the global agribusiness industry are not without contradictions outside of their tight relationship with nonrenewable fossil fuel energy. Many argue that peak oil (along with other resources such as water and phosphorous) will lead to peak food (Heinberg 2007; Pfeiffer 2006).[3] Additional contradictions include mono-cropping and the loss of biodiversity; cancerous petro-chemicals used to control pests and weeds; an epidemic of farmer suicides (most prevalent in India); mounting personal debt for farm inputs; soil erosion; the loss of local knowledge; the death of the small family farm and the corporate ownership, control and capitalization of food for profit. As important as these trends and contradictions are, perhaps the most globally stark has been the concomitant rise in obesity on the one hand and malnutrition on the other (Albritton 2009). Although the precise causes of the globalization of obesity are still debated by experts, one scholar has suggested that the petro-nutritional complex of high-calorie food combined with automobility are the major factors in ballooning body mass indexes (Roberts and Edwards 2010). According to the World Health Organization, this represents a major public health crisis: "Overweight and obesity are major risk factors for a number of chronic diseases, including diabetes, cardiovascular diseases and cancer. Once considered a problem only in high income countries, overweight and obesity are now dramatically on the rise in low- and middle-income countries, particularly in urban settings."[4]

Paralleling the epidemic of obesity, about a billion people remain undernourished and malnourished, which is a significant contributor to the deaths of 3.1 million children a year.[5] This is not because there is not enough food to feed the global population: "The growth of global agriculture's productive potential has so far been more than sufficient to exceed population growth, resulting in a steady, albeit slow, increase in average per capita food availability. For the world as a whole, per capita food availability has risen from about 2220 kcal/person/day in the early 1960s to 2790 kcal/person/day in 2006–08, while developing countries even recorded a leap from 1850 kcal/person/day to over 2640 kcal/person/day" (FAO 2012, 174).

The main problem, as Susan George (1976) identified many years ago, is not the availability of food but monetary access to food and productive land: the millions who go hungry, are malnourished or die of starvation do not have sufficient land to grow their own food or an income that would allow them to purchase an adequate diet on the market. In other words, the commodification of food and land has helped generate a situation where over 800 million people are undernourished despite decades of uneven progress in eradicating hunger (FAO 2014). This situation is only exacerbated by violent conflict, natural calamities such as droughts and floods, cuts in subsidies and state capacity due to debt and International Monetary Fund (IMF) economic restrictions and the liberalization of trade in food (Bello 2009; Patel and McMichael 2009, 10). If all this was not enough, mounting oil prices combined with the conversion of land to growing biofuels and financial speculation on the commodities market—what Russi (2013) calls the 'financialization of food'—seem to have pushed up food prices to record new levels, leading to a wave of food riots toward the end of the first decade of the twenty-first century (Baines 2014; Breger Bush 2012; Clapp 2014; Ghosh 2010; Isakson 2014).

Three researchers at the New England Complex System Institute have demonstrated that the wave of food riots across the Middle East, Asia and Africa during this period corresponded with spikes in the Food and Agricultural Organization's (FAO) Food Price Index (Lagi et al. 2011). They reason that riots are more likely when the index approaches or exceeds 210. Figure 6.0 charts the FAO's Food Price Index and the price of oil from 1961, when FAO data started to be recorded. Figure 6.1 shows the same data, but lags the food index by two years under the assumption that the increase in food prices may likely lag the price of oil because producers and retailers may not respond immediately by pushing up prices to consumers.

Both charts are revealing and suggest a strong correlation between oil prices and the FAO food index, particularly over the last ten years. Although correlation does not imply causation, it can hardly be doubted that oil prices have an effect on food prices, given that fossil fuels are involved in every node of the industrial agricultural system. Moreover, although it is difficult to tell with exact certainty, it is likely the case that prices are also being pushed up as more land is converted to the production of biofuels for cars and trucks

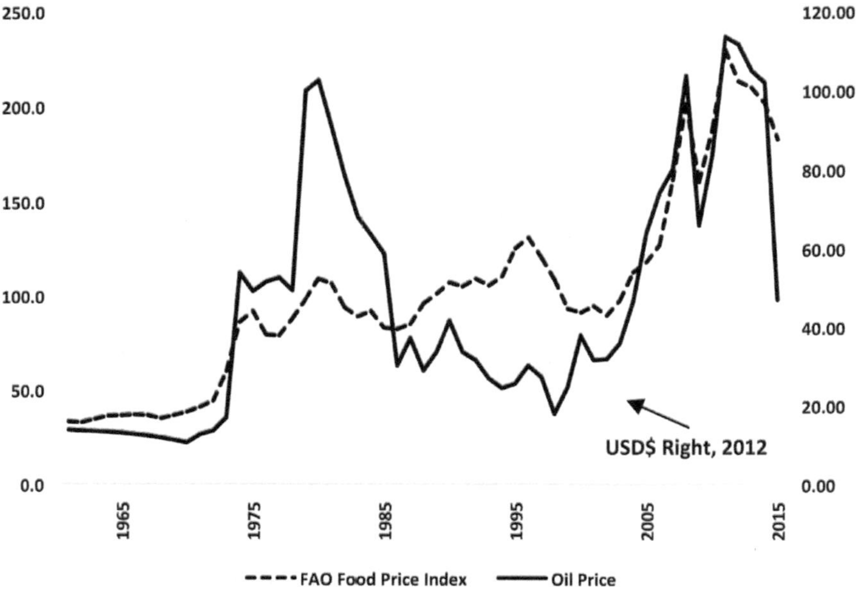

Figure 6.0. FAO Food Price and Oil Prices, 1961–2015. Source: FAO and BP Statistical Review and World Bank for 2013–15. Oil prices taken at January of each year.

Figure 6.1. FAO Food Price Index and Oil Prices, 1961–2015. Source: FAO and BP Statistical Review and World Bank for 2013–15. Oil prices taken at January of each year, Index 2 lag.

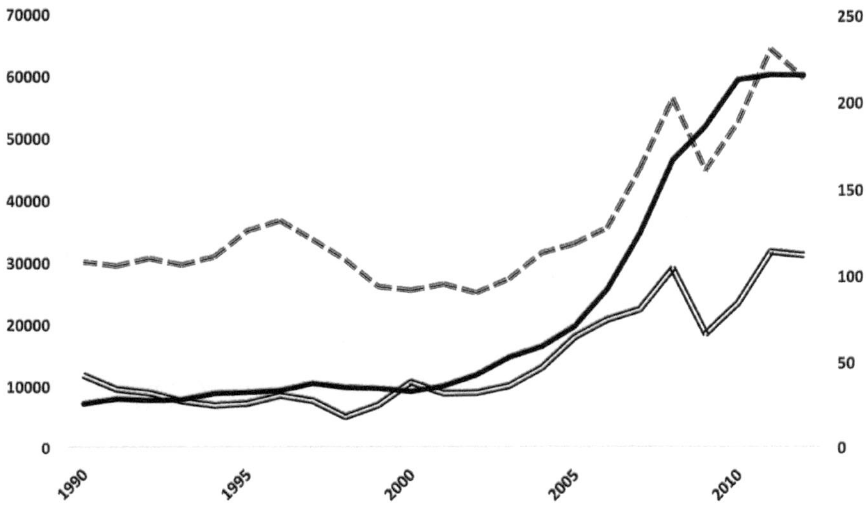

Figure 6.2. Biofuels Production (Ktoe), FAQ Food Price Index and Oil Price, 1990–2012. Source: BP Statistical Review and FAO. Note: Biofuel production, left (Ktoe), FAO Food Price Index and oil prices, right. The secondary axis doubles as 2012 US$ and the index figures for the FAO.

rather than food for humans. Figure 6.2 charts the considerable increase in biofuel production from 1990 with a clear upturn by 2005.

This spike was likely the result of giant government subsidies gifted to incentivize and stimulate farmers to produce biofuels. The production increase in fuel derived from organic matter also corresponds with increasing oil prices, as well as the visible increase in the FAO food index from 2004 or so. Going forward, these charts ultimately suggest a rather worrisome trend: elevated oil prices and the mass production of biofuels in the coming decades could push up food prices beyond the reach of many. Unless we can find a way to decarbonize and deindustrialize agriculture and disincentivize the production of biofuels, we are likely to see more pronounced food riots in the future, particularly among the most vulnerable who spend the majority of their small incomes on food. How governments and civil society will respond to the demands of food-insecure people is an open question, but it would be far better to restructure the food system along organic and local lines. Sadly, such a strategy runs counter to the corporate agri-food system and the short-term profits of the dominant owners of food. It will then be up to concerned members of civil society to act collectively to ensure their own food sovereignty, a process already begun by peasant farmers all over the globe.

The corporate control of food production, the inequality of access to nutritious food, land concentrated in the hands of the few and the general precarity of the corporate food regime have led many civil society actors to struggle for practical alternatives. Although some have argued for food security, defined by the World Food Summit in 1996 as a situation 'when all people at all times have access to sufficient, safe, nutritious food to maintain a healthy and active life', others believe that the more sound and sustainable approach is to pursue what La Via Campesina (The Peasants' Way) calls 'food sovereignty'.[6] Indeed, food security seems to be about whether there is enough to feed everyone, whereas food sovereignty is directly concerned with power and control over food regimes. Food sovereignty does not augur a simple definition, but its main idea is that communities of farmers and peasants should not only have the right to produce food, but also the right to define their own agricultural and food polices (McMichael 2014; Patel 2008, 302ff; 2009). This means that farmers and peasants rather than distant capitalized food corporations and their retailers are in greater control of their own food provisioning and thereby in greater control of their own livelihoods and destinies.

Although the struggles of the international peasant movement have made many advances and should be applauded, the portion of humanity caught up in the petro-nutritional agri-business complex should not delude itself with regard to the herculean task ahead. Creating an alternative food system that is fair, localized, sustainable and as organic as possible cannot be constructed overnight and will require entire communities to delink themselves from the neoliberal food regime and mobilize for their own versions and visions of food sovereignty (Pimentel and Pimentel 2008, 359; Roberts and Edwards 2010; Shiva 2008). If it is true that the 'whole structure and activities of [a] community are dependent upon questions of food supply', then transforming the global food system may be the only path to not only ensure populations against starvation, but also to ensure some level of future civility (Elton cited in Pimentel and Pimental 2008, 22).

Freshwater is another vital resource intimately connected with fossil fuels and the industrial food system. Water is essential for human health, biodiversity, industrial activity and agriculture (Schuster-Wallace and Sandford 2015). Only about 3 percent of water on earth is freshwater, with the remainder too saline for unprocessed human consumption. Of this tiny percentage, most of it is locked up in glaciers or in snow or is too deep underground for human use. It is, of course, true that saltwater can be desalinated, and there are more than 17,000 plants working across 150 countries to desalinate seawater. The International Desalination Association estimates that 300 million people rely on desalinated water for their health and well-being, or 4 percent of the global population.[7] It is estimated that as populations in more water-stressed regions of the globe experience shortages, a greater proportion of humanity may be served by desalination plants. This may be so, but at present the process is highly energy intensive and considerably more expensive than freshwater alternatives. Moreover, energy use is also required to deliver the water to private customers and industrial consumers. What this suggests is that industrial-scale desalination plants may not be sustainable over the long term as the energy inputs required for their functioning deplete and become more expensive over time.

Given this situation, can we then apply the concept of peak production to human water use? Two scholars suggest that we can, but they are cautious about how to define 'peak water' since 'water demonstrates characteristics of both renewable and nonrenewable resources' (Gleick and Palaniappan 2010, 11157). The authors suggest that when applied to water, peak water production can come in three forms. The first is what they call 'peak renewable water', and the concept applies to flow constraints per unit of time when resources are drawn—for example, from lakes—that exceed the rate of replenishment. The second is 'peak nonrenewable water', which is most evident in groundwater systems where production rates exceed the rate of replenishment over time or where contamination can lead to production declines. Last, the authors argue for a concept of 'peak ecological water' that calls our attention to a phenomenon whereby too much water is being diverted for human use, which causes such severe ecological damage that it exceeds the cost of accessing the water and jeopardizes the environment (Gleick and Palaniappan 2010). Gleick and Palaniappan make it clear that the world is not running out of freshwater resources. However, the various forms of 'peak water' are likely to be experienced in some communities more than others as the impacts of climate change, the intensive agricultural and industrial use of water and groundwater depletion start to mount and come into competition with nature and the need for ecological services. A 2015 report from the United Nations 'projected that the number of water-scarce countries could grow to 29 and the number of water-stressed countries is anticipated to rise to 19 by 2025. The combined population of these 48 countries is estimated to be 2.9 billion' (Schuster-Wallace and Sandford 2015, 27). Put another way, a significant portion of the global population could witness extreme precarity in access to one of the most fundamental resources needed to support human and natural life. The commercialization and privatization of water can exacerbate this problem insofar as water is held for the ransom of profit and those who cannot afford to buy water must go without or seek the assistance of civil society organizations or government agencies. Thus, peak water is fundamentally linked to the petro-market civilization constructed over centuries.

PEAK HEALTH, SCIENCE AND POPULATION

An underexplored consequence of higher prices for oil and gas is how this will affect the medical field and scientific knowledge. For example, most of the equipment used in modern medicine, from clinical disposals like syringes and plastic gloves to magnetic resonance imaging (MRI) and computed tomography (CT) scanners is fossil fuel dependent. Furthermore, the transport and energy infrastructure of modern hospitals is also heavily reliant on cheap gasoline since many modern hospitals are hardly in walking distance from suburban landscapes. Thus, how health care will be performed and delivered and at what cost in the future is just one more challenge that signals a general crisis of social reproduction. Indeed, there is some indication that public health officials have started to think about what peak oil portends for medical built environments and the geography of care (Hanlon and McGarney 2008).

But although we are likely to witness significant transformations in the medical field as energy becomes more expensive, we should also recall Jevons' claim that the scientific revolution also coincided with the extraction, production and consumption of coal. The discovery and industrial development of petroleum furthered scientific advancement, albeit in contradictory ways that contributed both to human betterment and the possibility of annihilating most species on the planet through nuclear war. Moreover, we should keep in mind that mass public education and the proliferation of universities that have been absolutely essential to advancing scientific knowledge are recent creations largely of the twentieth century despite some precursors. It could very well be that with increasing energy costs and declining energy availability, scientific knowledge begins to decline (Nikiforuk 2012, 158ff). Thus, since fossil fuels have been central to the globalization of modern medicine, equipment and scientific knowledge, it is likely that massive transformations in health, well-being and knowledge will be a key dimension of the post–carbon energy era and the universal crisis of social reproduction.

Another major, albeit sensitive, subject to consider is continued population growth amid finite resources. Indeed, as Ehrlich and Ehrlich remind us, concerns about human numbers should be 'related to the basic resources and ecosystem services required to support them' (2008, 141). More to the point, it is not so much the number of humans on the planet that is important, but their levels of differential consumption. By definition, poor people have a far smaller ecological footprint than do the world's 'high-consuming rich' who are unmistakably the real culprits in destroying the biosphere (Di Muzio 2015b; Erhlich and Ehrlich 2008, 156; Kempf 2008). What is also important to consider is that the population explosion that occurred from 1850 onward also corresponded with the greater exploitation of fossil fuels, scientific advancements and a more globalized and industrialized system of food provisioning designed around differential accumulation and the commodification of sustenance. For most, profit is the ransom for food in this system.

With these points at the forefront of our minds, it is worth considering that at the time of the agrarian revolution (about 10,000 years ago), the global population was estimated at 5 to 10 million, whereas today the United Nations Population Fund estimates there are 7 billion people on earth (Ehrlich and Ehrlich 2008, 143). These numbers are expected to increase by 2050, but no one knows with exact certainty what the global population will be. Yet even with increasing numbers, humanity's major problems remain the same: global climate change and the combustion of nonrenewable fossil fuels to sustain high-energy, high-mobility and high-consumption lifestyles for a privileged portion of humanity. These simple facts are often ignored by population alarmists who fetishize numbers rather than provide a critical investigation into how our world has been shaped and reshaped for the profit of the few and the consequences of the dominant form of imperial social reproduction (Baird 2011; Brand and Wissen 2013). But it is worth remembering, as McNeill points out, that 'although the twentieth century accounts for only 0.00025 of human history (100 out of 4 million years), it has hosted about a fifth of all human years' (2000, 9). A little reflection on this point demonstrates just how exceptional 'the little oil age' has been for a considerable portion of humanity. What dwindling supplies of fossil fuels and rising costs mean for the future of the global population is uncertain, but without profound change in the way we produce and consume—that is, in present patterns of social reproduction—there is a high chance that there will be far fewer of us around.

PEAK EMPLOYMENT AND PEAK MOBILITY

Rising energy costs will likely translate into higher prices for many goods and services, lower growth and mass unemployment. Affordable energy has supported an internationalized and extensive division *and* specialization of labor. As demand suffocates under higher energy prices and growth slows, unemployment is likely to increase, further depressing consumer demand. It will be difficult for economies to alleviate this vicious circle, as the previous bout of stagflation in the 1970s and early 1980s brought on by oil price spikes was not, as is commonly believed, alleviated by Volcker shocks, but by a dramatic decline in the cost of oil. Marx once noted that the 'production of too many useful things produces too large a useless population'.[8] In one sense, this hypothesis is tested daily in the hierarchical, gendered and racialized labour markets of the world as people are rendered unemployed or underemployed. With sustained triple-digit energy prices for oil, it is about to be tested on a far grander scale in the future. In the coming decades, what gets produced, how, where and by whom is an open question. It is highly likely that the forms of knowledge considered valuable will also change as more people may be drawn into producing basic necessities for subsistence as a first priority. As Hall and Day note, '[M]ost jobs would cease to exist without petroleum' (2009, 237). Patterns of mobility are likely to be transformed as well. Forms of social reproduction founded on gasoline-fired car culture and suburbanization will likely be significantly altered (Urry 2012). As the cost of overcoming distance mounts, long commutes to work, not to mention international business travel and transporting goods over long distances, will look increasingly unaffordable for individuals and businesses (Rubin 2009). Curtis (2009) has called this phenomenon 'peak globalization', and although it is not likely to happen any time soon, there is little doubt that the greater difficulty and expense of annihilating space with fossil fuel–powered mobility will change the constitution of the global political economy and force the relocalization of production and consumption on various communities. In addition, companies capitalized on the basis of selling petroleum-powered mobility, such as the recreational vehicle, aviation and the car industry, will watch their markets, workforce and market capitalization shrink as fewer people have the money to afford these forms of transport. Although it is impossible to foresee, suburban home values may also start to collapse as car-dependent suburban lifestyles appeal to fewer and fewer people. The only benefit of the suburbs may be that backyards or lawns could be converted into gardens for growing food, unlike in the cities where living space tends to be dense and access to soil at a premium. However, whether habitations in urban-built environments or suburban or rural environments will be prized in the future remains an open question. As previously mentioned, cities are highly energy intensive, and it is likely that urban forms of social reproduction will be transformed under the weight of higher energy costs for electricity, heating and cooling and the transport of food to supermarkets.

PEAK CLIMATE

The monetization and combustion of fossil fuels over at least the last three centuries has also had the effect of altering the climate. Despite the repeated attempts by Big Oil to confuse the public on the matter, the scientific consensus that human practices are leading to climate change is overwhelming and the likely effects on global social reproduction deadly serious (Klein 2014). In its latest report, the Intergovernmental Panel on Climate Change (IPCC 2014) notes that although regions will be affected in different ways, increasing temperatures will have incredible effects on human and natural ecosystems. With varying degrees of confidence, the report argues that should present trends continue, the world is likely to witness the disappearance of permafrost, melting glaciers, more severe weather events, shifting patterns of precipitation, drought, food shortages, stress on freshwater resources, more intense wildfires, the acidification of oceans, the loss of livelihoods and settlements, coastal erosion, reduced fisheries catch and urban floods, just to name some of the major estimates (IPCC 2014, 14).

Matters are more complex when we consider the far more underreported phenomenon of global dimming and brightening. From the 1950s, scientists observing pan evaporation noticed something strange in their time series data. Measuring pan evaporation is the leading way to assess the rate of water evaporation on the earth's surface, with radiation from the sun a leading cause of vaporization. In specific places, what scientists observed was that water was not evaporating as quickly as it had in the past, suggesting that less sunlight was hitting the surface of the earth. Scientists then hypothesized that although global 'dimming' could have natural causes, such as the eruption of a volcano, it also likely had human causes rooted in the pollution of the atmosphere. Under the guidance of Veerabhadran Ramanathan, a group of climate scientists performed an experiment in the Pacific, known as the Indian Ocean Experiment (INDOEX), to test the hypothesis. INDOEX was designed to study surface radiation in both the north and south islands of the Maldives. The air above the north islands is routinely polluted by India, whereas the southern islands receive clean air from Antarctica. What the team of scientists observed was that pollution particles ejected into the atmosphere from industry, automobility and the contrails of airplanes served to refract sunlight back into space, thus lessening the amount of solar energy entering the atmosphere in the polluted north islands. Pollution was effectively transforming clouds into giant mirrors reflecting radiation back into space.[9]

What this suggests is that global warming and climate change may be more threatening than scientists currently project, given that pollution particles are serving to help cool the planet by refracting some solar radiation back into space. Thus, if there is a precipitous drop in global air pollution, the climate may warm far quicker than currently expected due to global brightening (Wild 2009).

Although many remain hopeful that political pressure applied by civil society will help influence politicians across the world to come to some form of agreement that will keep temperatures from rising above two degrees, we can neither underestimate the power of the fossil fuel industry, nor can we discount differential accumulation and the political pursuit of growth as dominant logics that are not easily challenged or changed. It could very well be that what Gill has called 'the social reproduction of affluence' for the few trumps the future social reproduction of the most vulnerable, future generations and a relatively predictable climate (Bakker and Gill 2003). Moreover, the Pentagon and doubtless other state security agencies are actively preparing for the civil unrest both at home and abroad that may result from calamities associated with climate change, rising fuel and food prices or a general economic breakdown. As Ahmed argues, summing up a survey of recent Department of Defense (DoD) reports on the issue: "The Pentagon knows that environmental, economic and other crises could provoke widespread public anger toward government and corporations in coming years. The revelations on the NSA's global surveillance programmes are just the latest indication that as business as usual creates instability at home and abroad, and as disillusionment with the status quo escalates, Western publics are being increasingly viewed as potential enemies that must be policed by the state."[10]

What this suggests is that rather than dealing with the root causes of potential future calamities—the carbonization of everyday life and the logic of differential accumulation—the Pentagon appears to be preparing for the likelihood of mass civilizational disorder. In fact, compounding domestic threats due to climate change and food and fuel shortages may provide justifications for the greater militarization of society and a boon to weapons manufacturers.

PEAK CAPITALIZATION

Whereas skyrocketing energy costs may stuff the coffers of the international oil and gas companies in the decades to come, other sectors of the economy will likely watch their capitalization implode as oil supplies decline and costs rise. Although it is difficult to calculate with any precision, it is worth noting that the capitalization of *every* sector of the economy is interlinked and ultimately undergirded by affordable energy. Fossil fuels are not just a 'factor of production' that can be easily substituted as economists are wont to believe. For example, the 'mobility industry' could be said to be made up of four major sectors: (1) automobiles and parts, (2) aerospace and defence, (3) industrial transportation and (4) travel and leisure. Their total capitalization as of June 2014 was approximately US$2.6 trillion.[11] If we consider the market value of each sector, they are, respectively, ranked at sixth, twentieth, twenty-fourth and seventeenth out of thirty-seven sectors of the global economy. Since capitalization is a measure of the expectations investors have about the future, all we need to do is imagine a scenario where energy costs rise so high as to challenge the profit expectations of investors. In such a context, a run on these sectors of the economy is highly likely. We already have suggestive evidence for this process using the *Financial Times* Global 500. The market value of what I have called the mobility industry was US$1.5 trillion in 2007, up from US$1.3 trillion in 2006. But by 2009, the capitalization of this industry was almost halved to US$816 billion. With government bailouts of automotive firms and the reduction in the price of oil from its US$147 a barrel high in 2008, the capitalization of these four sectors has quite obviously recovered. But what of the capitalization of the leading 500 companies of the global economy during the record-breaking period of high oil prices of 2008? Within a year, global capitalization collapsed by 42 percent from about US$27 trillion to US$16 trillion by May 2009.[12]

The collapse of capitalization is conventionally blamed on the subprime mortgage crisis and its aftershocks. But as former Canadian Imperial Bank of Commerce (CIBC) economist Jeff Rubin (2008) has argued, this explanation of the crisis fails to account for the severity and widespread nature of the collapse in capitalization and the fact that recessionary conditions hit Japan and Europe *before* Wall Street imploded. According to Rubin, this suggests another culprit: the high price of oil as *the* key driver that enables or disables other sectors of the economy to meet the expectations of investors. And as he points out, there has been a strong historical correlation between high oil prices and the onset of major recessions (see also Hamilton 2009).

Perhaps we will never know what the precise cause of the global financial crisis was and we should be highly sceptical of explanations that single out one cause. However, given affordable energy's importance to every sector of the global economy, the high price of oil would seem a likely driver pushing global demand and profit expectations down for non-oil and gas firms. Moreover, I have shown that increases in capitalization coincided with increasing energy use. We would suspect then that as oil prices increase over time and growth begins to slow, the expectations of capitalists would diminish and, as a result, so would global capitalization. Thus, not only will rising energy prices have a considerable impact on differential accumulation as the oil and gas industry make record profits, but the process itself might also be the undoing of the entire edifice of global capitalization.

DEMOCRACY AND THE CIVIL PEACE

It is worthwhile to recall that democracy has never been the rule of human history, but rather its exception. But we should also recall that nowhere has there ever been an instantiation of democracy that is either complete or ideal. It seems that ever since the transition to agrarian communities, elite rule and appropriation have been a persistent norm in world order with democratic practices only serving to change the degree of elite rule and the amount of their appropriation from society. Braudel perhaps put it best when he observed that "conspicuous at the top of the pyramid is a handful of privileged people. Everything invariably falls into the lap of this tiny elite: power, wealth, a large share of surplus production . . . Is there not in short, whatever the society and whatever the period, an insidious law giving power to the few, an irritating law it must be said, since the reasons for it are not obvious. And yet this stubborn fact, taunting us at every turn. We cannot argue with it: all evidence agrees" (1983, 466).

How the few are able to dominate the many largely remains an unsolved mystery in the social sciences despite considerable attempts to theorize elite rule (e.g., Carr 1981, 46ff; Di Muzio 2015a; Gill 2008; Gramsci 1971; Mills 1956). In this conclusion we will get nowhere close to solving that mystery or weighing possible answers, but it is worth remembering that whatever the semblance of democracy we have today, a small portion of humanity—what I have called dominant owners—effectively receive most of the benefits of human activity through their ownership of income-generating assets and the capitalization of everyday life (Di Muzio 2015a; Di Muzio and Robbins 2015). Although scholars debate the precise causes and timing of democratization, it is largely a twentieth-century phenomenon both in its geographic expansion and as its relative deepening for the lower orders of humanity (Huntington 1993; Moore 1974; Potter et al. 2005; Roper 2013).

Despite the fact that democracy should be treated as a continual process and never a point of arrival, there is a relatively clear transition to more democratic forms of practice after World War II than in all human history put together. I do not intend to explore all the factors that have likely contributed to the process of democratization in various countries throughout the world. However, two factors seem to be absolutely crucial: education and rising standards of living connected to some kind of economic growth. But as we have hinted at, the greater proliferation of public education, not to mention higher education and economic growth, are largely contingent on the mass consumption of fossil fuel energy. In other words, the era of mass democratization, however imperfect, occurred during the carbonization of everyday life and, at least for many, the growing energy intensity of state–civil society complexes and an always unfinished liberal world order (Cox 1987; Latham 1997). I do not want to reduce decades of peace research to the one simple idea that civil peace within nations is achieved when the majority of a society can live reasonably well and where all the benefits and risks of modernity are shared relatively equally across society. But it would be counterintuitive to imagine that the reverse—a society of greater inequality and mass unemployment where there is greater everyday life precarity for the vast majority of citizens—is a recipe for a lasting civil peace both within and between nations.

Indeed, using examples from the twentieth century, Friedrichs' (2010, 2013) research has suggested that three responses to dwindling energy supplies and elevated prices by states could be (1) predatory militarism to take resources from others, (2) a form of totalitarian retrenchment where elite social forces protect their own privileges at the majority's expense and (3) a form of socioeconomic adaptation where government forces encourage and incentivize new forms of social reproduction less dependent on energy-intensive consumption and the imperatives of the market. Given the historical record, all of these practices could be considered likely as societies struggle to maintain a way of life that was from its inception unsustainable and unreproducible in the long term. How the transition to post-carbon national societies as well as a more generalized post-carbon world order is likely to be played out can only be approximated by theory and our knowledge of history, never fully predicted. But that the transition is inevitable and that future generations of humanity will undergo a general crisis of social reproduction can hardly be in doubt without widespread sustained social and economic change. The challenges of global warming only compound these problems, particularly for the most vulnerable. It is now time to conclude our study with a few final comments.

CODA

By any scale of human history, the little oil age within the broader age of carbon will be a short one. Many do not expect it to last much longer and many reason that that the twenty-first century will be more like Mad Max than the Jetsons. Still, there is little doubt that utopianism of one sort or another prevails in certain corners of scholarship and within the popular imagination of peoples across the world. Could technology or a new energy source keep humanity on a course of endless consumption and exponential growth and extend the benefits of a consumer society to an increasing population? Could the mounting global debt burden that has accompanied the rise of carbon capitalism ever truly be paid off? The weight of the evidence seems to suggest not and although hope may always spring eternal in the minds of optimists, so, too, might we say, do the laws of thermodynamics in the minds of scientists. These laws both enable and constrain possibilities for action, but as yet, cannot be negotiated with. But why, then, does humanity persist along such a ruinous course as if working daily toward the onset of a general crisis of social reproduction? Far be it from me to offer a definitive answer. But if I had to answer the question, if only in a tangential way, I would advance a two-fold thesis. First, I would adopt the Douglasian thesis that 'power concedes nothing without a demand'.[13] This is the plausible claim that power relations will continue to operate as they have been unless they are met with overwhelming resistance or something in the structural fabric of their power is significantly damaged, weakened or destroyed. The Wizard of Oz can continue being a wizard so long as no one peers behind the curtain. What I think this means in terms of counter-strategy is that not changing courses has to be made extremely costly—both in terms of morality and in terms of price—for those who want to continue upon the same immoral and ruinous path. In a world order governed by the logic of differential accumulation, outside of direct violence—which is not advocated for here—change must be promoted by making our rulers and the 1 percent they largely work for pay very high costs for their inaction, as well as the policies that keep us wedded to a politics of impending catastrophe.

Second, we may also do well to recognize what I will call the Carr–Diamond thesis. The thesis holds that definite upper-class forces can indeed benefit from actively harming others. This can happen directly or through structural forms of power and violence that systematically deny rights and privileges to a portion of humanity (Galtung 1969; Gill and Law 1988). For example, Carr argued that in international affairs 'the hollowness of the glib nineteenth-century platitude that nobody can benefit from what harms another was revealed' (1981, 58). Using national protectionist systems as his example, he wanted to show that pursuing the self-interest of the nation and achieving disproportionate benefits in a given context could have ruinous results for the community of nations as a whole. In a similar albeit distinct way, Jared Diamond also demonstrated how previous civilizational orders that had collapsed pursued a ruinous course of action because it benefited elites and enabled their social customs of symbolic power, at least for a time. Diamond put it thus at the conclusion of his study: "A further conflict of interest involving rational behavior arises when the interests of the decision-making elite in power clash with the interests of the rest of society. Especially if the elite can insulate themselves from the consequences of their actions, they are likely to do things that profit themselves, regardless of whether those actions hurt everybody else. All of these examples in the preceding several pages illustrate situations in which a society fails to try to solve perceived problems because the maintenance of the problem is good for some people" (2005, 431, 432).

Indeed, one of the main goals of our own study was to demonstrate how the constitution and reconstitution of forms of social reproduction and world order tied to fossil fuels were not inevitable, but socially constructed and reconstructed by the owners who came to dominate them, not to mention the owners of other sectors of the economy whose social reproduction and differential earnings became inextricably bound to affordable, abundant and accessible fossil fuels. Shaping this world order—albeit always against resistance—has indeed been 'good for some people', while the problems of carbon capitalism are increasingly pushed onto the majority of humanity and future generations, from increasing global debt burdens all the way to the pollution of lived environments and the catastrophes of climate change. In such a world order where the destruction of the biosphere continues to be monetized for the benefit of the 1 percent and affluent consumers, there is much work to do done by concerned citizens who do not want to 'draw the magic cap down over [their] eyes and ears as a make-believe that there are no monsters'. After all, it would be the ultimate tragedy of human development 'if the history of the human race proved to be nothing more noble than the story of an ape playing with a box of matches on a petrol dump' (Gore cited in Ehrlich and Ehrlich 2008, 158).

NOTES

1. https://www.marxists.org/history/erol/periodicals/spark/19470401.htm. From Marx's 1867 preface to *Capital*.
2. Maurice B. Green, 1978, *Eating Oil: Energy Use in Food Production* (Boulder: Westview Press) and Dale Allen Pfeiffer, 2006, *Eating Fossil Fuels: Oil, Food and the Coming Crisis in Agriculture* (Gabriola Island: New Society Publishers.
3. Tom Bawden, 'Have We Reached "Peak Food"? Shortages Loom as Global Production Rates Slow', *The Independent*, 28 January 2015.
4. World Health Organization, *Obesity* (http://www.who.int/topics/obesity/en/). See also Visscher and Seidell (2001).
5. World Food Programme, *Hunger Statistics* (http://www.wfp.org/hunger/stats).
6. http://www.who.int/trade/glossary/story028/en/.
7. http://idadesal.org/desalination-101/desalination-by-the-numbers/.
8. Karl Marx, *Economic and Philosophic Manuscripts of 1844* (http://www.marxists.org/archive/marx/works/1844/manuscripts/needs.htm).
9. http://www.bbc.co.uk/sn/tvradio/programmes/horizon/dimming_trans.shtml.
10. Nafeez Mosaddeq Ahmed, 'Pentagon Bracing for Public Dissent over Climate and Energy Shocks', *The Guardian*, 16 June 2013.
11. I use 2010 because in the last two *Financial Times* Global 500 reports, the sectors are disaggregated.
12. Anne-Britt Dullfo, 'FT Global 500 2009', *Financial Times*, 29 May 2009.
13. This quote is taken from Frederick Douglas' speech entitled 'West Indian Emancipation' in 1857 (http://www.blackpast.org/1857-frederick-douglass-if-there-no-struggle-there-no-progress).

Bibliography

Abouharb, M. Rodwan, and David Cingranelli. 2007. *Structural Adjustment and Human Rights*. Cambridge: Cambridge University Press.

Abramsky, Kolya. 2008–2009. "Energy and Labor in the World Economy." *The Commoner* 1 (13): 23–43.

———. 2010. *Sparking a Worldwide Energy Revolution: Social Struggles in the Transition to a Post-Petrol World*. Oakland, CA: AK Press.

Adelman, Morris. 1972–73. "Is the Oil Shortage Real? Oil Companies as OPEC Tax-Collectors." *Foreign Policy* 9: 69–107.

Agamben, Giorgio. 1998. *Homo Sacer: Sovereign Power and Bare Life*. Translated by Daniel Heller-Roazen. Palo Alto, CA: Stanford University Press.

Ahmed, Nafeez. 2013. "James Hansen: Fossil Fuel Addiction Could Trigger Runaway Global Warming." *Guardian, UK Blog*. http://www.theguardian.com/environment/earth-insight/2013/jul/10/james-hansen-fossil-fuels-runaway-global-warming.

Akins, James E. 1972. "The Oil Crisis: This Time the Wolf Is Here." *Foreign Affairs* 51:462–490.

Alam, M. Shahid. 2009. "Bringing Energy Back into Economy." *Review of Radical Political Economics* 41 (2): 170–185.

Albritton, Robert. 2000. *Let Them Eat Junk: How Capitalism Creates Hunger and Obesity*. London: Pluto Press.

———. 2009. *Let them Eat Junk: How Capitalism Creates Hunger and Obesity*. London: Pluto Press.

Alcott, Blake. 2005. "Jevons' Paradox." *Ecological Economics* 54 (1): 9–21.

Allen, Robert C. 1979. "International Competition in Iron and Steel, 1850–1913." *Journal of Economic History* 34 (4): 911–37.

———. 2011. "Why the Industrial Revolution Was British: Commerce, Induced Invention, and the Scientific Revolution." *Economic History Review* 64 (2): 357–84.

Altvater, Elmar. 2007. "The Social and Natural Environment of Fossil Capitalism." In *The Socialist Register 2007: Coming to Terms with Nature*, ed. Leo Panitch and Colin Leys. London: Merlin Press.

Anievas, Alexander, and Kerem Nisancioglu. 2013. "What's at Stake in the Transition Debate? Rethinking the Origins of Capitalism and the 'Rise of the West.'" *Millennium: Journal of International Studies* 42 (1): 78–102.

———. 2014. "The Poverty of Political Marxism." *International Socialist Review* 1 (94): 1–25.

Anonymous. 2011. "History of Energy Consumption in the United States, 1775–2009." *Oil & Gas Journal* 109 (14): SA11.

Appleby, Joyce Oldham. 2001. "The Vexed Story of Capitalism Told by American Historians." *Journal of the Early Republic* 21 (1): 1–18.

Arrighi, Giovanni. 1994. *The Long Twentieth Century.* London: Verso.

———. 1997. *Adam Smith in Bejing*. London: Verso.

Ashworth, John. 2011. "Towards a Bourgeois Revolution? Explaining the American Civil War," *Historical Materialism* 19 (4): 193-205.

Ashworth, William, and Mark Pegg. 1986. *The History of the British Coal Industry. Volume 5: 1946–1982: The Nationalized Industry*. Oxford: Oxford University Press.

Aspromourgos, Tony. 2005. "The Invention of the Concept of Social Surplus: Petty in the Hartlib Circle." *The European Journal of the History of Economic Thought* 12 (1): 1-24.

Assadourian, Erik. 2010. "The Rise and Fall of Consumer Cultures." *State of the World 2010: Transforming Cultures, from Consumerism to Sustainability.* Worldwatch Institute. http://www.worldwatch.org/files/pdf/Chapter%201.pdf.

Auty, Richard. 1993. *Sustaining Development in Mineral Economies: The Resource Curse Thesis*. London: Routledge.

Ayres R. U., L. W. Ayres, and B. Warr. 2003. "Exergy, Power and Work in the US Economy, 1900–1998." *Energy* 28 (3): 219–73.

Bae, Christopher J., Wei Wang, Jianxin Zhao, Shengming Huang, Feng Tian, and Guanjun Shen. 2014. "Modern Human Teeth from Late Pleistocene Luna Cave (Guangxi, China)." *Quaternary International*, http://dx.doi.org/10.1016/j.quaint.2014.06.051, 1–15.

Baines, Joseph. 2014. "Food Price Inflation as Redistribution: Towards a New Analysis of Corporate Power in the World Food System." *New Political Economy* 19 (1): 79–112.

Baird, Vanessa. 2011. *The No Nonsense Guide to World Population.* Oxford: New Internationalist.

Bakker, Isa, and Stephen Gill, eds. 2003. *Power, Production and Social Reproduction*. Basingstoke: Palgrave.

Balaam, David N., and Bradford Dillman. 2013. *Introduction to International Political Economy*. 6th edition. New York: Routledge.

Banner, Stuart. 2005. *How the Indians Lost their Land: Law and Power on the Frontier*. Cambridge, MA: Belknap Press.

Baptist, Edward E. 2014. *The Half has Never Been Told: Slavery and the Making of American Capitalism*. New York: Basic Books.

Barca, Stefania. 2011. "Energy, Property and the Industrial Revolution Narrative." *Ecological Economics* 70 (7): 1309–15.

Bardi, Ugo. 2001. *The Limits to Growth Revisited.* Florentino: Springer.

———. 2009. "Peak Oil: The Four Stages of a New Idea." *Energy* 34:323–26.

Barnes, Robert. 1972. "International Oil Companies Confront Governments: A Half-Century of Experience." *International Studies Quarterly* 16 (4): 454–71.

Bartolovich, Crystal. 2010. "A Natural History of Food Riots." *New Formations* 1 (69): 42–61.

Bas, Muhummet A., and Andrew J. Coe. 2012. "Arms Diffusion and War." *Journal of Conflict Resolution* 56 (4): 651–74.

Baskin, Jonathan Barron, and Paul J. Miranti Jr. 1997. *A History of Corporate Finance*. Cambridge: Cambridge University Press.

Beckert, Sven. 2004. "Emancipation and Empire: Reconstructing the Worldwide Web of Cotton Production in the Age of the American Civil War." *The American Historical Review* 109 (5): 1405-1438.

Beaud, Michel. 2001. *A History of Capitalism*, 1500–2000. New York: Monthly Review Press.

Behravesh, Maysam. 2012. "The Formative Years of Anglo-Iranian Relations (1907–1953): Colonial Scramble for Iran and Its Political Legacy." *Digest of Middle East Studies* 21 (2): 386–400.

Beker, Avi. 1982. "The Arms-Oil Connection: Fuelling the Arms Race." *Armed Forces & Society* 8 (3): 419–42.

Bell, Stephanie. 2000. "Do Taxes and Bonds Finance Government Spending?" *Journal of Economic Issues* 34 (3): 603–20.

Bello, Walden. 2009. *Food Wars*. London: Verso.

Benjamin, Walter. 1968. *Illuminations*. Edited by Hannah Arendt. New York: Harcourt, Brace & World.

Bensel, Richard Franklin. 1990. *Yankee Leviathan: The Origins of Central State Authority in America, 1859–1877*. Cambridge: Cambridge University Press.

Berlin, Ira. 1998. *Many Thousands Gone: The First Two Centuries of Slavery in America*. Cambridge, MA: Belknap Press.

Bernstein, Peter L. 2004. *The Power of Gold: The History of an Obsession*. New York: John Wiley and Sons.

Bhambra, Gurminder K. 2009. *Rethinking Modernity: Postcolonialism and the Sociological Imagination*. Basingstoke: Palgrave Macmillan.

———. 2010. "Historical Sociology, International Relations and Connected Histories." *Cambridge Review of International Affairs* 23 (1): 127–43.

Bina, Cyrus. 1988. "Internationalization of the Oil Industry: Simple Oil Shocks or Structural Crisis." *Review* 11 (3): 329-370.

———. 2006. "The Globalization of Oil," *International Journal Of Political Economy* 35 (2): 4-34.

Black, Brian. 2000. *Petrolia: Creating the North American Landscape*. Baltimore: John Hopkins University Press.

Blackburn, Robin. 2010. *The Making of New World Slavery: From the Baroque to the Modern, 1492–1800.* London: Verso.

———. 2011. *The Overthrow of Colonial Slavery, 1776–1848*. London: Verso.

Blum, William. 2004. *Killing Hope: U.S. Military and CIA Interventions since World War II*. Monroe: Common Courage Press.

Blyth, Mark, ed. 2009. *Routledge Handbook of International Political Economy.* London: Routledge.

Boehm, Christopher. 2001. *Hierarchy in the Forest: The Evolution of Egalitarian Behavior*. Cambridge, MA: Harvard University Press.

Bouton, Terry. 2001. "Review: Whose Original Intent? Expanding the Concept of the Founders." *Law and History Review* 19 (3): 661–71.

Brand, Ulrich, and Markus Wissen. 2013. "Crisis and Continuity of Capitalist Society-Nature Relationships: The Imperial Mode of Living and the Limits to Environmental Governance." *Review of International Political Economy* 20 (4): 687–711.

Braudel, Fernand. 1983. *The Wheels of Commerce: Civilization and Capitalism 15th to 18th Century.* Translated by Sian Reynolds. London: William Collins Sons & Co. Ltd.

Braun, Timothy F., and Lisa M. Glidden. 2014. *Understanding Energy and Energy Policy*. London: Zed.

Breen, T. H. 1985. *Tobacco Culture: The Mentality of the Great Tidewater Planters on the Eve of Revolution*. Princeton, NJ: Princeton University Press.

Breger Bush, Sarah. 2012. *Derivatives and Development: A Political Economy of Global Finance, Farming, and Poverty*. New York: Palgrave Macmillan.

Brennan, Jordan. 2012. "The Power Underpinnings, and Some Distributional Consequences, of Trade and Investment Liberalization in Canada." *New Political Economy* 18 (5): 715-747.

Brenner, Robert. 1976. "Agrarian Class Structure and Economic Development in Pre-Industrial Europe." *Past & Present* 7: 30-75.

———. 1977. "The Origins of Capitalist Development: a Critique of Neo-Smithian Marxism" *New Left Review* 104 (July-August): 25-92.

———. 1978. "Dobb on the Transition from Feudalism to Capitalism" *Cambridge Journal of Economics* 2: 121-140.

Brewer, John. 1989. *The Sinews of Power: War, Money and the English State: 1688-1783*. London: Unwin Hyman.

Bromley, Simon. 1991. *American Hegemony and World Oil: The Industry, the State System and the World Economy.* Cambridge: Polity Press.

———. 2005. "The United States and the Control of World Oil." *Government and Opposition* 40 (2): 225–55.

Brown, Christopher. 1973. "International Political Economy: Some Problems of an Inter-Disciplinary Enterprise." *International Affairs* 49 (1): 51–60.

Brown, E. Cary. 1989. *Episodes in the Public Debt History of the United States.* Working Paper, Department of Economics, MIT, No. 540.

Bruff, Ian. 2012. "Authoritarian Neoliberalism, the Occupy Movements, and IPE." *Journal of Critical Globalisation Studies* 5:114–16.

Bruff, Ian, and Daniela Tepe. 2011. "What is Critical IPE?" *Journal of International Relations and Development* 14:354–58.

Bunker, Stephen G., and Paul S. Ciccantell. 2005. *Globalization and the Race for Resources*. Baltimore: John Hopkins University Press.

Burke, Sharon E. 2014. "Powering the Pentagon: Creating a Lean, Clean Fighting Machine." *Foreign Affairs* 93 (3): 33–37.

Byers, Terence J. 1996. "Differentiation of the Peasantry Under Feudalism and the Transition to Capitalism: In Defence of Rodney Hilton." *Journal of Agrarian Change* 6: 17–68.

Campbell, Colin J. 2002. "Petroleum and People." *Population and Environment* 24 (2): 193–207.

Campbell, David. 2005. "The Biopolitics of Security: Oil, Empire and the Sport Utility Vehicle." *American Quarterly* 57 (3): 943–72.

Campbell, John. 1977. "Oil Power in the Middle East." *Foreign Affairs* 56 (1): 89–110.

Cannan, Edwin. 1921. "Early History of the Term Capital." *The Quarterly Journal of Economics* 35 (3): 469–81.

Canny, Nicholas. 1998. *The Oxford History of the British Empire: Volume I: The Origins of Empire: British Overseas Enterprise to the Close of the Seventeenth Century*. Oxford: Oxford University Press.

Capgemini and RBC Wealth Management. *World Wealth Report* 2013.

Carr, E. H. 1981. *The Twenty Years' Crisis, 1919–1939*. Basingstoke: Palgrave.

Cassis, Yousel. 2006. *Capitals of Capital: A History of International Financial Centres, 1780–2005*. Cambridge: Cambridge University Press.

Chiao, R. Y., M. L. Cohen, A. J. Leggett, W. D. Phillips, and C. L. Harper Jr., eds. 2010. *Visions of Discovery: New Light on Physics, Cosmology, and Consciousness*. Cambridge: Cambridge University Press.

Chilingarian, George V., and Teh Fu Yen, eds. 1978. *Bitumens, Asphalts and Tar Sands*. New York: Elsevier Scientific Publishing Company.

Chomsky, Noam, and Edward S. Herman. 1979. *A Political Economy of Human Rights*. Montreal: Black Rose Books.

Chubin, S. 1976. "The International Politics of the Persian Gulf." *British Journal of International Studies* 2 (3): 216–30.

Cipolla, Carol M. 1977. *Before the Industrial Revolution: European Society and Economy, 1000-1700*. Third Edition. London: Routledge.

Clapp, Jennifer. 2014. "Financialization, Distance, and Global Food Politics." *The Journal of Peasant Studies* 41 (5): 797–814.

Clark, Brett, and Richard York. 2005. "Carbon Metabolism: Global Capitalism, Climate Change and the Biospheric Rift." *Theory and Society* 34:391–428.

Clark, Brett, Andrew K. Jorgenson, and Jeffrey Kentor. 2010. "Militarization and Energy Consumption." *International Journal of Sociology* 40 (2): 23–43.

Clark, William R. 2005. *Petrodollar Warfare: Oil, Iraq, and the Future of the Dollar*. Gabriola Island: New Society Publishers.

Clay, C. G. A. 1984. *Economic Expansion and Social Change: England 1500-1700*. Two volumes. Cambridge UK: Cambridge University Press.

Cleveland, Harold van B., and W. H. Bruce Brittain. 1975. "A World Depression?" *Foreign Affairs* 53 (2): 223–41.

Cochrane, Thomas C. 1961. "Did the Civil War Retard Industrialization?" *The Mississippi Valley Historical Review* 48 (2): 197–210.

Cohn, Theodore M. 2011. *Global Political Economy*. Sixth Edition. London: Pearson.

Colgan, Jeff D. 2013a. "Fueling the Fire: Pathways from Oil to War." *International Security* 38 (2): 147–80.

———. 2013b. *Petro-Aggression: When Oil Causes War*. Cambridge: Cambridge University Press.

Coll, Steve. 2012. *Private Empire: ExxonMobil and American Power*. New York: Penguin Group.

Collingham, Lizzie. 2012. *The Taste of War: World War Two and the Battle for Food*. London: Penguin Books.

Commons, John R. 1959. *Legal Foundations of Capitalism*. Madison: University of Wisconsin Press.

Cowen, David Jack. 2000. *The Origins and Economic Impact of the First Bank of the United States, 1791–1797*. New York: Garland Publishing.

Cox, Robert W. 1981. "Social Forces, States and World Orders." *Millennium: Journal of International Studies* 10 (2): 126–55.

———. 1987. *Production, Power, and World Order*. New York and London: Columbia University Press.

———. 2007. "The International in Evolution." *Millennium: Journal of International Studies* 35 (3): 513–27.

Cromar, Peter. 1977. "The Coal Industry on Tyneside, 1771–1800: Oligopoly and Spatial Change." *Economic Geography* 53 (1): 79–94.

Crosby, Alfred. 2006. *Children of the Sun: A History of Humanity's Unappeasable Appetite for Energy*. New York: W. W. Norton.

Curtis, Fred. 2009. "Peak Globalization: Climate Change, Oil Depletion, and Global Trade." *Ecological Economics* 69 (2): 427–34.

Curtis, Thomas D. 2014. "Editor's Introduction: Updating the Curtis Thesis' and Riches, Real Estate, and Resistance: How Land Speculation, Debt, and Trade Monopolies Led to the American Revolution." *American Journal of Economics and Sociology* 73: 445–626.

Custers, Peter. 2010. "Military Keynesianism Today: An Innovative Discourse." *Race & Class* 51 (4): 79–94.

Davies, Glyn. 2002. *A History of Money: From Ancient Times to the Present Day*. Cardiff: University of Wales Press.

Davis, Clarence, Kenneth E. Wilburn Jr., and Ronald E. Robinson. 1991. *Railway Imperialism*. New York: Greenwood Press.

Davis, David Brion. 2006. *Inhuman Bondage: The Rise and Fall of Slavery in the New World*. Oxford: Oxford University Press.

Dearne, Martin J., and Keith Branigan. 1995. "The Use of Coal in Roman Britain." *The Antiquaries Journal* 75:71–105.

Debeir, Jean-Claude, Jean-Paul Deléage, and Daniel Hémery. 1991. *In the Servitude of Power: Energy and Civilization through the Ages*. London: Zed Books.

Deffeyes, Kenneth. 2001. *Hubbert's Peak: The Impending World Oil Shortage*. Princeton, NJ: Princeton University Press.

———. 2005. *Beyond Oil: The View from Hubbert's Peak*. New York: Hill and Wang.

de Graaff, Naná. 2012. "The Hybridization of the State-Capital Nexus in the Global Energy Order." *Globalizations* 9 (4): 531–45.

Denning, Michael. 2010. "Wageless Life." *New Left Review* 66:79–97.

De Soto, Hernando. 2004. *The Mystery of Capital: Why Capitalism Triumphs in the West and Fails Everywhere Else*. New York: Basic Books.

DeZeeuw, J. W. 1978. "Peat and the Dutch Golden Age." *AAG Bijdragen* 21:3–31.
Diamond, Jared. 1997. *Guns, Germs, and Steel: The Fates of Human Societies*. New York: W. W. Norton.
———. 2005. *Collapse: How Societies Choose to Fail or Succeed*. New York: Penguin Books.
Dickins, Amanda. 2006. "The Evolution of International Political Economy." *International Affairs* 82 (3): 479–92.
Dimmock, Spencer. 2014. *The Origin of Capitalism in England, 1400–1600*. Leiden: Brill.
Di Muzio, Tim. 2007. "The Art of Colonization: The Capitalization of the State and the Ongoing Nature of Primitive Accumulation." New Political Economy 12 (4): 517–539.
———. 2008. "Governing Global Slums: The Biopolitics of Target 11." Global Governance 14: 305-326.
———. 2011. "The Crisis of Petro-Market Civilization: The Past as Prologue?" In *Global Crises and the Crisis of Global Leadership*, ed. Stephen Gill, 73–88. Cambridge: Cambridge University Press.
———. 2012. "Capitalizing a Future Unsustainable: Finance, Energy and the Fate of Market Civilization." *Review of International Political Economy* 19 (3): 363–88.
———, ed. 2014. *The Capitalist Mode of Power: Engaging the Power Theory of Value*. London and New York: Routledge.
———. 2015a. *The 1% and the Rest of Us: A Political Economy of Dominant Ownership*. London: Zed Books.
———. 2015b. "The Plutonomy of the 1%: Dominant Ownership and Conspicuous Consumption in the New Gilded Age." *Millenium: Journal of International Studies*, 43 (2): 492–510.
Di Muzio, Tim, and Richard Robbins. 2015. *Debt as Power*. London: Bloomsbury.
Dobb, Maurice. 1946. Reprinted 1963. *Studies in the Development of Capitalism*. London: Routledge and Kegan Paul.
Draper, Nicholas. 2007. "Possessing Slaves": Ownership, Compensation and Metropolitan Society in Britain at the Time of the Emancipation 1834–40." *History Workshop Journal* 1 (64): 75–102.
———. 2010. *The Price of Emancipation: Slave-Ownership, Compensation and British Society at the End of Slavery*. Cambridge: Cambridge University Press.
Drescher, Seymour, and Stanley Engerman. 1998. *A Historical Guide to World Slavery*. Oxford: Oxford University Press.
Duncan, T. K., and Christopher J. Coyne. 2013. "The Origins of the Permanent War Economy." *The Independent Review* 18 (2): 219–40.
Dyer, Christopher. 1997. "The Howard Linecar Lecture 1997. Peasants and Coins: The Uses of Money in the Middle Ages." *British Numismatic Journal* 67:30–47.
Edling, Max M., and Mark D. Kaplanoff. 2004. "Alexander Hamilton's Fiscal Reform: Transforming the Structure of Taxation in the Early Republic." *William and Mary Quarterly* 61 (4): 713–44.
Ehrlich, Paul R., and Anne H. Ehrlich. 2008. *The Dominant Animal: Human Evolution and the Environment*. Washington, DC: Island Press.
EIA. 2014. *Canada: Overview*. http://www.eia.gov/countries/cab.cfm?fips=ca.

Eichengreen, Barry. 2008. *Globalizing Capital: A History of the International Monetary System*. Princeton, NJ: Princeton University Press.
Einhorn, Robin L. 2006. *American Taxation, American Slavery*. Chicago: University of Chicago Press.
El-Gamal, Mahmoud A., and Amy Myers Jaffe. 2010. *Oil, Dollars, Debt, and Crisis: The Global Curse of Black Gold*. Cambridge: Cambridge University Press.
Elhefnawy, Nader. 2008. "The Impending Oil Shock." *Survival* 50 (2): 37–66.
Energy Watch Group. 2007. *Coal: Resources and Future Production*. March. http://www.energywatchgroup.org/fileadmin/global/pdf/EWG_Report_Coal_10-07-2007ms.pdf.
Engdahl, William. 2004. *A Century of War: Anglo-American Oil Politics and the New World Order*. London: Pluto Press.
———. 2012. *Myths, Lies and Oil Wars.* Weisaden: edition.engdahl.
Engels, Jeremy. 2005. "Reading the Riot Act: Rhetoric, Psychology, and Counter-Revolutionary Discourse in Shays's Rebellion, 1786–1787." *Quarterly Journal of Speech* 91 (1): 63–88.
Evans, Peter, Dietrich Rueschneyer, and Theda Skocpol, eds. 1985. *Bringing the State Back In*. Cambridge: Cambridge University Press.
ExxonMobil. 2014. *The Outlook for Energy*. http://corporate.exxonmobil.com/en/energy/energy-outlook (7/15/2015).
FAO. 2012. *Food and Agricultural Organization: Statistical Yearbook, 2012*. Rome. http://www.fao.org/docrep/015/i2490e/i2490e00.htm.
———. 2014. *State of Food Insecurity in the World, 2014*. Rome. http://www.fao.org/publications/sofi/2014/en/.
Fenn, Elizabeth A. 2000. "Biological Warfare in Eighteenth-century North America." *Journal of American History* 86:552–58.
Ferguson, E. James. 1954. "Speculation in the Revolutionary Debt: The Ownership of Public Securities in Maryland, 1790." *The Journal of Economic History* 14:35–45.
———. 1979. *The American Revolution: A General History, 1763–1790.* Homewood: The Dorsey Press.
Ferguson, Niall. 2006. *The Cash Nexus: Money and Power in the Modern World, 1700–2000.* New York: Basic Books.
———. 2011. *Civilization: The West and the Rest*. New York: Penguin.
———. 2012. *Empire: How Britain Made the Modern World*. New York: Penguin.
Fernandez-Armesto, Felipe. 2000. *Civilizations*. London: Macmillan.
———. 2001. *Food: A History*. London: Macmillan.
Fine, Ben. 1988. "The British Coal Industry's Contribution to the Political Economy of Paul Sweezy." *History of Political Economy* 20 (2): 235–50.
Flavin, Francis E. 2002. "A Pox on Amherst: Smallpox, Sir Jeffery, and a Town Named Amherst." *Historical Journal of Massachusetts* 30 (1): 1–29.
Foner, Eric. 2011. "The Civil War and Slavery: A Response." *Historical Materialism* 19 (4): 92-98.
Foster, John Bellamy. 1999. "Marx's Theory of Metabolic Rift: Classical Foundations for Environmental Sociology." *American Journal of Sociology* 105 (2): 366–405.
Foster, John Bellamy, and Hannah Holleman. 2010. "The Financial Power Elite." *Monthly Review* 62 (1): 1–17.

Foucault, Michel. 1982. "The Subject and Power." *Critical Inquiry* 8 (4): 777–95.
———. 1984. *The Foucault Reader.* Edited by Paul Rabinow. New York: Pantheon Books.
Fouquet, Roger, and Peter J. G. Pearson. 1998. "A Thousand Years of Energy Use in the United Kingdom." *The Energy Journal* 19 (4): 1–41.
Fresia, Jerry. 1988. *Toward an American Revolution: Exposing the Constitution and Other Illusions*. Boston: South End Press.
Frieden, Jeffry A., and David A. Lake. 2003. *International Political Economy: Perspectives on Global Power and Wealth*. Fourth edition. London: Routledge.
Friedenberg, Daniel M. 1992. *Life, Liberty and the Pursuit of Land: The Plunder of Early America*. Buffalo: Prometheus Books.
Friedrichs, Jörg. 2010. "Global Energy Crunch: How Different Parts of the World Would React to a Peak Oil Scenario." *Energy Policy* 38:4562–69.
———. 2011. "Peak Energy and Climate Change: The Double Bind of Post-normal Science." *Futures* 43:469–77.
———. 2013 *The Future is Not What it Used to Be: Climate Change and Energy Scarcity.* Cambridge, MA: MIT Press.
Frymer, Paul. 2014. "'A Rush and a Push and the Land Is Ours': Territorial Expansion, Land Policy, and U.S. State Formation." *Perspectives on Politics* 12 (1): 119-144.
Galtung, Johan. 1969. "Violence, Peace, and Peace Research." *Journal of Peace Research* 6 (3): 167–91.
Garnett, George. 2009. *The Norman Conquest: A Very Short Introduction*. Oxford: Oxford University Press.
Geisst, Charles R. 2000. *Monopolies in America: From Jay Cooke to Bill Gates*. Oxford: Oxford University Press.
George, Susan. 1976. *How the Other Half Dies: The Real Reasons for World Hunger*. New York: Penguin.
———. 1988. *A Fate Worse than Debt*. London: Penguin.
Georgescu-Roegen, Nicholas. 1971. *The Entropy Law and Economic Process*. Cambridge MA: Harvard University Press.
———. 1976. *Energy and Economic Myths: Institutional and Analytical Economic Essays*. New York: Pergamon Press.
Ghosh, Jayati. 2010. "The Unnatural Coupling: Food and Global Finance." *Journal of Agrarian Change* 10 (1): 72–86.
Gilens, Martin, and Benjamin I. Page. 2014. "Testing Theories of American Politics: Elites, Interest Groups, and Average Citizens." *Perspectives on Politics* 12 (3): 564–81.
Gill, Stephen. 1991. *American Hegemony and the Trilateral Commission*. Cambridge: Cambridge University Press.
———. 1995. "Globalization, Market Civilization and Disciplinary Neoliberalism." *Millennium: Journal of International Studies* 24 (3): 399–423.
———. 2008. *Power and Resistance in the New World Order.* Basingstoke: Palgrave.
———, ed. 2011. *Global Crises and the Crisis of Global Leadership.* Cambridge: Cambridge University Press.

Gill, Stephen. 1991. American Hegemony and the Trilateral Commission. Cambridge: Cambridge University Press.

Gill, Stephen, and David Law. 1988. *The Global Political Economy: Perspectives, Problems and Policies*. Baltimore: Johns Hopkins University Press.

Gilpin, Robert. 1975. *U.S. Power and the Multinational Corporation: The Political Economy of Foreign Direct Investment.* New York: Basic Books.

———. 2001. *Global Political Economy: Understanding the International Economic Order*. Princeton, NJ: Princeton University Press.

Gleick, Peter H., and Meena Palaniappan. 2010. "Peak Water Limits to Freshwater Withdrawal and Use." *Proceedings of the National Academy of Sciences of the United States of America* 107 (25): 11155–62.

Goldrick, James. 2014. "Coal and the Advent of the First World War at Sea." *War in History* 21 (3): 322–37.

Goldstone, Jack A. 2002. "Efflorescences and Economic Growth in World History: Rethinking the 'Rise of the West" and the Industrial Revolution." *Journal of World History* 13 (2): 323–89.

Gowan, Peter. 1999. *The Global Gamble: Washington's Bid for Global Dominance*. London Verso.

Gramsci, Antonio. 1971. *Selections from the Prison Notebooks*. Edited and translated by Quintin Hoare and Geoffrey Nowell Smith. New York: International Publishers.

Green, Maurice B. 1978. *Eating Oil: Energy Use in Food Production. Boulder*: Westview Press.

Greene, Jack P., and Richard M. Jellison. 1961. "The Currency Act of 1764 in Imperial-Colonial Relations, 1764–1776." *The William and Mary Quarterly* 18 (4): 485–518.

Grimmett, Richard F. 2004. "Instances of Use of United States Armed Forces Abroad, 1798–2004." Foreign Affairs, Defense, and Trade Division. Washington, DC: Congressional Research Service Library of Congress, report RL30172.

Grove, Eric J. 2005. *The Royal Navy since 1815: A New Short History*. Basingstoke: Palgrave MacMillan.

Hager, Sandy Brian. 2013a. "What Happened to the Bondholding Class? Public Debt, Power and the Top One Per Cent." *New Political Economy* 19 (2): 155–82.

———. 2013b. "America's Real 'Debt Dilemma'" *Review of Capital as Power*, 1, (1), 41–62.

Hall, Charles A. S., and John W. Day. 2009. "Revisiting the Limits to Growth." *American Scientist* 97 (3): 230–37.

Hall, Charles A. S., and Kent A. Klitgaard. 2012. *Energy and the Wealth of Nations: Understanding the Biophysical Economy*. New York: Springer.

Hall, Peter A., and David Soskice. 2001. *Varieties of Capitalism: The Institutional Foundations of Comparative Advantage*. Oxford: Oxford University Press.

Hallwood, Paul, and Stuart Sinclair. 1982. "OPEC's Developing Relationships with the Third World." *International Affairs* 58 (2): 271–86.

Hamilton, Clive. 2004. *Growth Fetish.* London: Pluto Press.

Hamilton, James D. 1983. "Oil and the Macroeconomy since World War II." *Journal of Political Economy* 91 (2): 228–48.

———. 2009. "Causes and Consequences of the Oil Shock of 2007–8." *Brookings Papers on Economic Activity* (Spring): 215–61, http://www.brookings.edu/~/media/Files/Programs/ES/BPEA/2009_spring_bpea_papers/2009_spring_bpea_hamilton.pdf.

Hancock, Kathleen J., and Vlado Vivoda. 2014. "International Political Economy: A Field Born of the OPEC Crisis Returns to Its Energy Roots." *Energy Research & Social Science*, 1:206–16.

Hanlon, P., and G. McGarney. 2008. "Peak Oil: Will It be Public Health's Greatest Challenge?" *Public Health* 122 (7): 647–52.

Hansen, James, Makiko Sato, Gary Russell, and Pushker Kharecha. 2013. "Climate Sensitivity, Sea Level and Atmospheric Carbon Dioxide." *Philosophical Transactions of the Royal Society Association* 371:1–31.

Harman, Chris. 2004. "The Rise of Capitalism." *International Socialism* (2nd series) 102 (Spring: np). https://www.marxists.org/archive/harman/2004/xx/risecap.htm (7/15/2015)

Harman, Chris, and Robert Brenner. 2004. "The Origins of Capitalism." *International Socialism* 2 (111):127–62.

———. 2006. "The Origins of Capitalism." *International Socialism* (2nd series), 111 (Summer: np). https://www.marxists.org/archive/harman/2006/xx/wbrenner.htm (7/15/2015)

Hartung, William D. 1955. "U.S. Conventional Arms Transfers: Promoting Stability or Fueling Conflict?" *Arms Control Today* 25 (9): 9–13.

Hartung, William. 1995. *And Weapons for All*. New York: HarperCollins.

Harvey, A. D. 2012. "Was the American Civil War the First Modern War?" *History* 97:272–80.

Harvey, David. 2003. *The New Imperialism*. Oxford: Oxford University Press.

Healy, Denis. 1979. "Oil, Money, Recession." *Foreign Affairs* 58 (2): 217–30.

Heinberg, Richard. 2003. *The Party's Over: Oil, War and the Fate of Industrial Societies.* Gabriola Island: New Society Publishers.

———. 2007. *Peak Everything: Waking Up to a Century of Declines.* Gabriola Island: New Society Publishers.

———. 2009. *Searching for a Miracle: Net Energy Limits and the Fate of Industrial Society*. Forum on Globalization and Post-Carbon Institute. http://www.postcarbon.org/report/44377-searching-for-a-miracle.

———. 2011. *The End of Growth: Adapting to Our New Economic Reality.* Gabriola Island: New Society Publishers.

Heinberg, Richard, and Daniel Lerch, eds. 2010. *The Post Carbon Reader: Managing the 21st Century's Sustainability Crises*. Healdsburg, CA: Watershed Media.

Heller, Henry. 1985. "The Transition Debate in Historical Perspective." *Science & Society* 49 (2): 208-213.

———. 2011. *The Birth of Capitalism: A Twenty-first Century Perspective.* London: Pluto Press.

Henle, Robert J. 1980. "Alcohol Fuels: Ford vs. Rockefeller." *America* 142 (3): 60-62.

Henry, James S. 2003. *The Blood Bankers*. New York: Four Walls Eight Windows.

Henwood, Doug. 1997. *Wall Street: How it Works and for Whom*. London: Verso.

Herman, Arthur. 2005. "The 'Myth' of British Seapower." *Orbis* 49 (2): 337–52.

Heuman, Gad, and Trevor Burnard. 2011. *Routledge History of Slavery*. London: Routledge.

Higgins, Matthew, Thomas Klitgaard, and Robert Lerman. 2006. "Recycling Petrodollars." *Current Issues: Federal Reserve Bank of New York* 12 (9): 1-7.

Hill, Christopher. 1959. *The English Revolution, 1640*. London: Lawrence & Wishart.

———. 1985. *The English Revolution*, 1640. London: Lawrence & Wishart.

Hilton, Rodney. 1976a. "Feudalism and the Origins of Capitalism." *History Workshop Journal* 1 (1): 9-25.

———. ed. 1976b. *Peasants, Knights and Heretics: Studies in Medieval English Social History*. Cambridge UK: Cambridge University Press.

Hiro, Dilip. 2006. *Blood of the Earth: The Battle for the World's Vanishing Oil Resources*. New York: Nation Books.

Hirsch, Robert L., Roger Bezdek, and Robert Wendling. 2005. *Peaking of World Oil Production: Impacts, Mitigation, & Risk Management*. http://www.netl.doe.gov/publications/others/pdf/oil_peaking_netl.pdf.

Hirst, Paul, and Grahame Thomas. 1996. *Globalization in Question: The International Economy and the Possibilities of Governance*. London: Polity.

Hobson, John M. 2004. *The Eastern Origins of Western Civilization*. Cambridge UK: Cambridge University Press.

———. 2013a. "Part 1—Revealing the Eurocentric Foundations of IPE: A Critical Historiography of the Discipline from the Classical to the Modern Era." *Review of International Political Economy* 20 (5): 1024–54.

———. 2013b. "Part 2—Reconstructing the Non-Eurocentric Foundations of IPE: From Eurocentric 'Open Economy Politics' to Inter-civilizational Political Economy." *Review of International Political Economy* 20 (5): 1055–81.

Holton, Woody. 1999. *Forced Founders: Indians, Debtors, Slaves and the Making of the American Revolution in Virginia.* Chapel Hill: University of North Carolina Press.

———. 2005a. "An 'Excess of Democracy'—or a Shortage?" *Journal of the Early Republic* 25 (3): 339–82.

———. 2005b. "Divide et Impera: Federalist 10 in a Wider Sphere." *The William and Mary Quarterly* 62 (2): 175–212.

———. 2005c. "Did Democracy Cause the Recession That Led to the Constitution?" *The Journal of American History* 92 (2): 442–69.

Hoppit, Julian. 2002. "The Myths of the South Sea Bubble." Transactions of the Royal Historical Society 12: 141-165.

Huber, Matthew T. 2008. "Energizing Historical Materialism: Fossil Fuels, Space and the Capitalist Mode of Production." *Geoforum* 40 (1): 105–15.

———. 2013. *Lifeblood: Oil, Freedom and the Forces of Capital.* Minneapolis: University of Minnesota Press.

Hudson, Michael. 2005. [1977]. *Global Fracture: The New International Economic Order.* London: Pluto Press.

Hülsemeyer, Axel. 2010. *International Political Economy: A Reader*. Don Mills, ON: Oxford University Press.

Humphrey, William S., and Joe Stanislaw. 1979. "Economic Growth and Energy Consumption in the UK, 1700–1975." *Energy Policy* 7 (1): 29–42.

Hunter, Robert. 1919. *Violence and the Labor Movement*. New York: The Macmillan Company.
Huntington, Samuel. 1993. *The Third Wave: Democratization in the Late 20th Century*. Norman: University of Oklahoma Press.
Huston, James L. 1993. "The American Revolutionaries, the Political Economy of Aristocracy, and the American Concept of the Distribution of Wealth, 1765–1900." *The American Historical Review* 98 (4): 1079–1105.
Hybel, Alex Roberto. 2012. *Ideology in World Politics: From the Roman Empire to Al-Qaeda*. London: Routledge.
IEA. 2010. *World Energy Outlook: Press and Media*. http://www.worldenergyoutlook.org/pressmedia/quotes/42/index.html.
———. 2012. *Response System for Oil Supply Emergencies*. https://www.iea.org/publications/freepublications/publication/EPPD_Brochure_English_2012_02.pdf.
———. 2013. *Key World Energy Statistics*. Paris, France.
———. 2014a. *World Energy Outlook*. France.
———. 2014b. *World Energy Investment Outlook*. France.
———. 2014c. *Renewable Energy Medium Term Market Report*. France.
Ingham, Geoffrey. 2004. *The Nature of Money*. Cambridge: Polity Press.
Inglehart, Ronald, and Christian Welzel. 2005. *Modernization, Cultural Change and Democracy: The Human Development Sequence*. Cambridge: Cambridge University Press.
IPCC. 2014. *Climate Change 2014: Synthesis Report*. Summary for Policymakers. http://www.ipcc.ch/report/ar5/syr/.
Isakson, S. Ryan. 2014. "Food and Finance: The Financial Transformation of Agro-food Supply Chains." *The Journal of Peasant Studies* 41 (5): 749–75.
Jabber, Paul. 1978. "Conflict and Cooperation in OPEC: Prospects for the Next Decade." *International Organization* 32 (2): 377–99.
Jackson, Andrew, and Ben Dyson. 2013. *Modernizing Money: How our Monetary System is Broken and How It Can be Fixed*. London: Positive Money.
Jackson, Tim. 2009. *Prosperity without Growth: Economics for a Finite Planet*. London: Earthscan.
Jensen, Merrill. 1958. *The New Nation: A History of the United States during the Confederation*. New York: Alfred A. Knopf.
Jeremy, David. 1977. "Damming the Flood: British Government Efforts to Check the Outflow of Technicians and Machinery, 1780–1843." *Business History Review* 51 (1): 1–34.
Jevons, William Stanley. 1865. *The Coal Question: An Inquiry Concerning the Progress of the Nation, and the Probable Exhaustion of Our Coal-Mines*. London: MacMillan and Co.
Jones, Geoffrey. 1981. *The State and the Emergence of the British Oil Industry*. London: Macmillan.
———. 2000. *Merchants to Multinationals: British Companies in the Nineteenth and Twentieth Centuries*. Oxford: Oxford University Press.
———. 2005. *Multinationals and Global Capitalism: From the Nineteenth to the Twenty-first Century*. Oxford: Oxford University Press.

Jones, R. J. Barry. 1981a. “International Political Economy: Problems and Issues—Part I.” *Review of International Studies* 7 (4): 245–60.

———. 1981b. “International Political Economy: Problems and Issues—Part II.” *Review of International Studies* 8 (1): 39–52.

Josephson, Matthew. 1934. *The Robber Barrons*. New York: Harcourt, Brace and Company.

Jurado-Sánchez, José. 2009. “Military Expenditure, Spending Capacity and Budget Constraint in Eighteenth-Century Spain and Britain.” *Revista de Historia Económica* 27 (1): 141–74.

Karl, Terry Lynn. 1997. *The Paradox of Plenty*. Berkeley: University of California Press.

Kempf, Hervé. 2008. *How the Rich are Destroying the Earth*. Translated by Leslie Thatcher. Foxhole, UK: Green Books.

Kennedy, Geoff. 2006. “Digger Radicalism and Agrarian Capitalism.” *Historical Materialism* 14 (3): 113–43.

———. 2008. *Diggers, Levellers and Agrarian Capitalism: Radical Political Thought in Seventeenth-Century England*. Lanham, MD: Lexington Books.

Kennedy, Paul. 1987. *The Rise and Fall of the Great Powers: Economic Change and Military Conflict from 1500 to 2000*. London: Unwin Hyman.

Kent, Marian. 1976. *Oil and Empire: British Policy and Mesopotamian Oil 1900–1920*. London: Macmillan.

Keohane, Robert. 1984. *After Hegemony: Cooperation and Discord in the World Political Economy*. Princeton, NJ: Princeton University Press.

Kindleberger, Charles, and Robert Z. Aliber. 2005. *Manias, Panics and Crashes: A History of Financial Crises*. Basingstoke: Palgrave MacMillan.

Kinsella, David. 1998. “Arms Transfer Dependence and Foreign Policy Conflict.” *Journal of Peace Research* 35 (1): 7–23.

Klare, Michael. 2002. *Resource Wars: The New Landscape of Global Conflict*. New York: MacMillan.

———. 2004. *Blood and Oil: The Dangers and Consequences of America's Growing Dependency on Imported Petroleum*. New York: Metropolitan Books.

———. 2009. *Rising Powers, Shrinking Planet: The New Geopolitics of Energy*. New York: Holt.

———. 2012. *The Race For What's Left: The Global Scramble For The World's Last Resources*. New York : Metropolitan Books.

———. 2013. “The Third Carbon Age.” http://www.tomdispatch.com/blog/175734/.

Klein, Naomi. 2014. *This Changes Everything: Capitalism vs. the Climate*. Toronto: Alfred A. Knopf.

Knollenberg, Bernhard. 1954. “General Amherst and Germ Warfare.” *The Mississippi Valley Historical Review* 41:489–94.

Knowles, Helen J. 2007. “The Constitution and Slavery: A Special Relationship.” *Slavery & Abolition: A Journal of Slave and Post-Slave Studies* 28 (3): 309–28.

Konings, Martijn. 2011. *The Development of American Finance*. New York: Cambridge University Press.

Kulikoff, Allan. 1992. *The Agrarian Origins of American Capitalism*. Charlottesville: University Press of Virginia.

Kunstler, James Howard. 2005. *The Long Emergency: Surviving the End of Oil, Climate Change and Other Converging Catastrophes of the Twenty-First Century.* New York: Grove Press.

Kwarteng, Kwasi. 2014. *War and Gold: A 500-Year History of Empires, Adventures and Debt.* Public Affairs.

Labban, Mazen. 2008. *Space, Oil and Capital.* London: Routledge.

Lagi, Marco, Karla Z. Bertrand, and Yaneer Bar-Yam. 2011. "The Food Crises and Political Instability in North Africa and the Middle East." New England Complex Systems Institute. http://necsi.edu/research/social/food_crises.pdf.

Landes, David. 1998. *The Wealth and Poverty of Nations: Why Some Are So Rich and Some Are So Poor.* New York: W. W. & Norton Company.

Latham, Robert. 1997. *The Liberal Moment: Modernity, Security, and the Making of Postwar International Order.* New York: Columbia University Press.

Lawrence, K. S. 2008. "The Origins of Capitalism and the 'Rise of the West.'" *International Journal Of Comparative Sociology* 49 (6): 501-5.

Le Billon, Philippe. 2005. *Geopolitics of Resource Wars: Resource Dependence, Governance and Violence.* London: Frank Cass.

Lefebvre, Georges, et. al. 1982 *The Transition from Feudalism to Capitalism.* London: Verso.

Levinson, Marc. 2006. *The Box: How the Shipping Container Made the World Smaller and the World Economy Bigger.* Princeton, NJ: Princeton University Press.

Levy, Hermann. 1911. [2001]. *Monopoly and Competition: A Study in English Industrial Organization.* Kitchener: Batoche Books.

Levy, Michael B. 1983. "Freedom, Property and the Levellers: The Case of John Lilburne." *The Western Political Quarterly* 36 (1): 116–33.

Levy, W. J. 1979. "Oil and the Decline of the West." *Foreign Affairs* 58 (5): 999–1015.

Lewis, J. O. 1921. "America's Petroleum Problem." *Journal of the Franklin Institute* 191 (3): 357–79.

Lewis, John P. 1974. "Oil Other Scarcities, and the Poor Countries." *World Politic* 27 (1): 63–86.

Li, Minqi. 2007. "Peak Oil, the Rise of China and India and the Global Energy Crisis." *Journal of Contemporary Asia* 74 (4): 449–71.

———. 2008. *The Rise of China and the Demise of the Capitalist World-Economy.* London: Pluto Press.

Lieber, Robert J. 1979. "Europe and America in the World Energy Crisis." *International Affairs* 55 (4): 531–45.

Linebaugh, Peter. 1976. "Karl Marx, the Theft of Wood, and Working Class Composition: A Contribution to the Current Debate." *Crime and Social Justice* 6:5–16.

Lloyd, T. O. 1984. *The British Empire 1558–1983.* Oxford: Oxford University Press.

Luxemburg, Rosa. 1913. [1951]. *The Accumulation of Capital.* Translated by Agnes Schwarzschild. New Haven, CT: Yale University Press.

Malanima, Paolo. 2006. "Energy Crisis and Growth 1650–1850: The European Deviation in a Comparative Perspective." *Journal of Global History* 1:101–21.

Magdoff, Fred, and John Bellamy Foster. 2011. *What Every Environmentalist Needs to Know about Capitalism.* New York: Monthly Review Press.

Malm, Andreas. 2012. "China as Chimney of the World: The Fossil Capitalism Hypothesis." *Organization and the Environment*, 25 (2): 146-77.

———. 2013. "The Origins of Fossil Capital: From Water to Steam in the British Cotton Industry." *Historical Materialism* 21 (1): 15–68.

Mankiw, N. Gregory. 2009. *Macroeconomics.* Seventh edition. New York: Worth Publishers.

Mann, Barbara Alice. 2005. *George Washington's War on Native America.* Westport, CT: Praeger.

Mann, Bruce H. 2003. *Republic of Debtors: Bankruptcy in the Age of American Independence.* Cambridge, MA: Harvard University Press.

Mann, Charles C. 2013. "What If We Never Run Out of Oil." *The Atlantic* 311 (4): 48–63.

Manning, Richard. 2004. "The Oil We Eat." *Harper's* (February): 37–45.

Marx, Karl. 1887. *Capital: A Critique of Political Economy Volume 1.* Translated by Samuel Moore and Edward Aveling. Moscow: Progress Publishers.

———. 1894/1996. *Capital, Volume III.* New York: International Publishers. https://www.marxists.org/archive/marx/works/1894-c3/ (7/15/2015).

———. 1976. *Capital: Volume 1.* London: Penguin.

Marx, Karl, and Fredrick Engels. 1848. *Manifesto of the Communist Party.* Translated by Samuel Moore. https://www.marxists.org/archive/marx/works/1848/communist-manifesto/ (7/15/2015).

———. 1861. *Writings on the North American Civil War.* http://www.marxists.org/archive/marx/works/download/Marx_Engels_Writings_on_the_North_American_Civil_War.pdf.

Maugeri, Leonardo. 2006. *The Age of Oil.* Westport, CT: Praeger.

McKie, Robin. 2013. "Why Did the Neanderthals Die Out." *The Observer*, June 2.

McKillop, Andrew, and Sheila Newman, eds. 2005. *The Final Energy Crisis.* London: Pluto Press.

McKinsey Global Institute. 2013. *Financial Globalization: Retreat or Reset.* Global Capital Markets Report.

McMahon, James. 2013. "The Rise of a Confident Hollywood: Risk and the Capitalization of Cinema." *Review of Capital as Power* 1 (1): 23-40.

McMichael, Philip. 2009. "A Food Regime Analysis of the World Food Crisis." *Agriculture and Human Values* 4:281–95.

———. 2014. "Historicizing Food Sovereignty." *The Journal of Peasant Studies* 41 (6): 933–57.

McNeill, J. R. 2000. *Something New Under the Sun: An Environmental History of the Twentieth-Century World.* New York: W. W. Norton Company.

McPherson, James M. 2007. *This Mighty Scourge: Perspectives on the Civil War.* Oxford: Oxford University Press.

Meadows, Donella, and Dennis Meadows. 1972. *The Limits to Growth: A Report for the Club of Rome's Project on the Predicament of Mankind.* London: Potomac Associates.

Mellars, Paul. 2006. "Why Did Modern Human Populations Disperse from Africa ca. 60,000 Years Ago: A New Model." *Proceedings of the National Academy of Sciences USA* 103 (25): 9381–86.

Melman, Seymour. 1970. *Pentagon Capitalism: The Political Economy of War*. New York: McGraw-Hill.

———. 1974. *The Permanent War Economy: American Capitalism in Decline*. New York: Simon and Shuster.

———. 1997. "From Private to State Capitalism: How the Permanent War Economy Transformed the Institutions of American Capitalism." *Journal of Economic Issues* 31 (2): 311–30.

Michie, Ranald. 2006. *The Global Securities Market: A History*. Oxford UK, Oxford University Press.

Mielants, Eric H. 2007. *The Origins of Capitalism and the Rise of the West*. Philadelphia: Temple University Press.

Mikadashi, Zuhayr. 1980. "Oil Exporting Countries and Oil-Importing Countries: What Kind of Interdependence?" *Millennium: Journal of International Studies* 9 (1): 1–20.

———. 1981. "Oil Prices and OPEC Surpluses: Some Reflections." *International Affairs* 57 (3): 407–27.

Miller, Raymond C. 2008. *International Political Economy: Contrasting Views*. New York: Routledge.

Mills, C. Wright. 1956. [2000]. *The Power Elite*. Oxford: Oxford University Press.

Milonakis, Dimitris, and Ben Fine. 2009. *From Political Economy to Economics Method, the Social and the Historical in the Evolution of Economic Theory*. London: Routledge.

Mintz, Sidney W. 1986. *Sweetness and Power: The Place of Sugar in Modern History*. London: Penguin Books.

Mirowski, Phillip. 1989. *More Heat Than Light: Economics as Social Physics, Physics as Nature's Economics*. New York: Cambridge University Press.

Mitchell, Timothy. 2011. *Carbon Democracy: Political Power in the Age of Oil*. London: Verso.

Moe, E. Mancur. 2009. "Olson and Structural Change." *Review of International Political Economy* 16 (2): 202–30.

Moe, Espen. 2010. "Energy, Industry and Politics: Energy, Vested Interests, and Long-term Economic Growth and Development." *Energy* 35:1730–40.

Mohr, Anton. 1926. *The Oil War*. New York: Harcourt, Brace and Company.

Moore, Barrington Jr. 1974. *The Social Origins of Dictatorship and Democracy: Lord and Peasant in the Making of the Modern World*. New York: Penguin University Books.

Moore, Jason W. 2003. "Nature and the Transition from Feudalism to Capitalism." *Review* 26 (2): 97–172.

Morales, Alex. 2014. "Oil Investors May be Running off a Cliff They Can't See." *Bloomberg*, December 2.

Morgan, Edmund S. 1957. "The American Revolution: Revisions in Need of Revising." *The William and Mary Quarterly* 14 (1): 3–15.

Morgan, K. 2001. "Slavery and the Debate over Ratification of the United States Constitution." *Slavery & Abolition: A Journal of Slave and Post-Slave Studies* 22 (3): 40–65.

Morris, Richard B. 1962. "Class Struggle and the American Revolution." *The William and Mary Quarterly* 19 (1): 3-29.

Moss, Michael. 2014. *Salt, Sugar, Fat: How the Food Giants Hooked Us*. New York: Random House.

Mulligan, Shane. 2010. "Reassessing the Crisis: Ecology and Liberal International Relations." *Alternatives* 35:137–62.

Mumford, Lewis. 1934. *Technics and Civilization*. London: Routledge and Kegan Paul.

Murphy, Craig. 1984. *The Emergence of the NIEO Ideology*. Boulder, CO: Westview Press.

Neal, Larry. 1990. *The Rise of Financial Capitalism: International Capital Markets in the Age of Reason*. Cambridge: Cambridge University Press.

Neeson, J. M. 1993. *Commoners: Common Right, Enclosure and Social Change in England 1700–1820*. Cambridge: Cambridge University Press.

Nef, John, U. 1966. [1932]. *The Rise of the British Coal Industry, Volumes I–II.* Abingdon: Frank Cass.

———. 1977. "An Early Energy Crisis and Its Consequences." *Scientific American* (November): 140–50.

Newell, Peter, and Mathew Paterson. 2010. *Climate Capitalism: Global Warming and the Transformation of the Global Economy*. Cambridge: Cambridge University Press.

Nikiforuk, Andrew. 2010. *Dirty Oil and the Future of a Continent*. Vancouver: Greystone Books.

———. 2012. *The Energy of Slaves: Oil and the New Servitude*. Vancouver: Greystone Books.

Nitzan, Jonathan, and Shimshon Bichler. 1995. "Bringing Capital Accumulation Back In: The Weapondollar-Petrodollar Coalition-Military Contractors, Oil Companies and Middle East 'Energy Conflicts.'" *Review of International Political Economy* 2 (3): 446–515.

———. 2002. *The Global Political Economy of Israel*. London: Pluto Press.

———. 2004. *A Political Economy of Israel*. London: Pluto Press.

———. 2009. *Capital as Power: A Study of Order and Creorder*. London: Routledge.

———. 2015. "Still About Oil?" *Real-World Economics Review I* (70): 49-79.

Nore, Peter, and Terisa Turner, eds. 1980. *Oil and Class Struggle*. London: Zed.

Noreng, Oystein. 2007. *Crude Power: Politics and the Oil Market*. New York: I. B. Tauris and Co. Ltd.

Nye, David E. 1999. *Consuming Power: A Social History of American Energies.* Cambridge MASS: MIT Press.

Oatley, Thomas. 2011. *International Political Economy: Interests and Institutions in the Global Economy*. Fifth edition. New York: Pearson.

O'Brien, Robert, and Marc Williams. 2010. *Global Political Economy: Evolution and Dynamics*. Third edition. New York: Palgrave.

Ohline, Howard A. 1971. "Republicanism and Slavery: Origins of the Three-Fifths Clause in the United States Constitution." *The William and Mary Quarterly* 28 (4): 563–84.

Oppenheim, V. H. 1976–1977. "Why Oil Prices Go Up: The Past: We Pushed Them Up." *Foreign Policy* 23:24–57.

Ovadia, Jesse Salah. 2013. "The Making of Oil-Backed Indigenous Capitalism in Nigeria." *New Political Economy* 18 (2): 258–83.

———. 2014. *The Petro-Developmental State in Africa: Making Oil Work in Angola, Nigeria and the Gulf of Guinea*. London: Hurst.

Overbeek, Henk. 1990. *Global Capitalism and National Decline: The Thatcher Decade in Perspective*. London: Unwin Hyman.

Overton, Mark. 1996a. "Re-establishing the English Agricultural Revolution." *The Agricultural History Review* 44 (1): 1–20.

———. 1996b. *Agricultural Revolution in England: The Transformation of the Agrarian Economy 1500–1850*. Cambridge: Cambridge University Press.

Painter, David S. 1986. *Oil and the American Century: The Political Economy of U.S. Foreign Oil Policy, 1941–1954*. Baltimore: John Hopkins University Press.

Palan, Ronen, ed. 2000. *Global Political Economy: Contemporary Theories*. London: Routledge.

Panitch, Leo, and Sam Gindin. 2012. *The Making of Global Capitalism: The Political Economy of American Empire*. London: Verso.

Parra, Francisco. 2004. *Oil Politics: A Modern History of Petroleum*. London: I. B. Tauris.

Patel, Raj. 2008. *Stuffed and Starved: From Farm to Fork, the Hidden Battle for the World's Food System*. London: Portobello Books.

———. 2009. "Food Sovereignty." *Journal of Peasant Studies* 36 (3): 663–706.

Patel, Raj, and Philip McMichael. 2009. "A Political Economy of the Food Riot" review. *Fernand Braudel Center* 32 (1): 9–35.

Penrose, Edith. 1979. "OPEC's Importance in the World Oil Industry." *International Affairs* 55 (1): 18–32.

Perelman, Michael. 2000. *The Invention of Capitalism: Classical Political Economy and the Secret History of Primitive Accumulation*. Durham, NC: Duke University Press.

Perkins, John. 2004. *Confessions Of An Economic Hit Man*. San Francisco: Berrett-Koehler Publishers.

Perrow, Charles. 2002. *Organizing America: Wealth, Power and the Origins of Corporate Capitalism*. Princeton, NJ: Princeton University Press.

Pfeiffer, Dale Allen. 2006. *Oil, Food and the Coming Crisis in Agriculture*. Gabriola Island: New Society Publishers.

Philips, Nicola, ed. 2005. *Globalizing International Political Economy*. Basingstoke: Palgrave Macmillan.

Phillips, Kevin. 2006. *American Theocracy: The Peril and Politics of Radical Religion, Oil and Borrowed Money in the 21st Century*. New York: Viking.

Pimental, David, and Marcia H. Pimentel. 2008. *Food, Energy and Society*. Third edition. London: CRC Press.

Podobnik, Bruce. 2006. *Global Energy Shifts: Fostering Sustainability in a Turbulent Age*. Philadelphia: Temple University Press.

Polanyi, Karl. 1957. The Great Transformation: *The Political and Economic Origins of our Times. Boston*: Beacon Press.

Polimeni, John, Kozo Mayumi, Mario Giampietro, and Blake Alcott. 2008. *The Jevons Paradox and the Myth of Resource Efficiency Improvements*. Abingdon: Earthscan.

Pollack, Gerald A. 1974. "The Economic Consequences of the Energy Crisis." *Foreign Affairs* 52 (3): 452–71.

Pollack, Sheldon David. 2009. *War, Revenue, And State Building: Financing The Development of The American State*. Ithaca, NY: Cornell University Press.

Porter, Eduardo. 2015. "Behind Drop in Oil Prices, Washington's Hand." *New York Times*, January 20.

Post, Charles. 2012. *The American Road to Capitalism: Studies in Class-Structure, Economic Development and Political Conflict, 1620–1877*. Chicago: Haymarket Books.

———. 2014. "The American Path of Bourgeoisie Development Revisited: A Response." *Science & Society* 78 (4): 369–79.

Potter, David S., David Goldblatt, Margaret Kiloh, and Paul Lewis. 2005. *Democratization*. Milton Keynes: Polity Press.

Prak, Maarten, ed. 2001. *Early Modern Capitalism: Economic and Social Change in Europe 1400–1800*. London: Routledge.

Price, T. Douglas. 1995. "Social Inequality at the Origins of Agriculture." In *Foundations of Social Inequality*, ed. Gary M. Feinman and T. Douglas Price, 129–51. London: Plenum Press.

Przeworski, Adam, et al. 2000. *Democracy and Development: Political Institutions and Well-Being in the World, 1950–1990*. Cambridge: Cambridge University Press.

Raval, Angli. 2015. "The Big Drop: Riyadh's Oil Gamble." *Financial Times*. 9 March 2015.

Ravenhill, John, ed. 2008. *Global Political Economy*. Second edition. Oxford: Oxford University Press.

Reid, John Phillip. 1988. *The Concept of Liberty in the Age of the American Revolution*. Chicago: The University of Chicago Press.

Robbins, Lionel. [1935]. 1949. *An Essay on the Nature and Significance of Economic Science*. Second edition. London: Macmillan.

Roberts, Ian, with Phil Edwards. 2010. *The Energy Glut: The Politics of Fatness in an Overheating World*. London: Zed.

Robinson, William I. 2004. *A Theory of Global Capitalism: Production, Class, and State in a Transnational World*. Baltimore: Johns Hopkins University Press.

Rodrigue, Jean-Paul, Claude Comtois, and Brian Slack. 2006. *The Geography of Transport Systems*. London: Routledge.

Roper, Brian S. 2013. *A History of Democracy: A Marxist Interpretation*. London: Pluto Press.

Rothbard, Murray N. 2002. *A History of Money and Banking in the United States: The Colonial Era to World War II*. Auburn: Ludwig von Mises Institute.

Rothman, Adam. 2005. *Slave Country: American Expansion and the Origins of the Deep South*. Cambridge, MA: Harvard University Press.

Rowbotham, Michael. 1998. *The Grip of Death: A Study of Modern Money, Debt Slavery and Destructive Economics*. Charlbury: Jon Carpenter Publishing.

Roy, William G. 1997. *Socializing Capital: The Rise of the Industrial Corporation in America*. Princeton, NJ: Princeton University Press.

Rubin, Jeff. 2008. *Why Your World is About to Get a Whole Lot Smaller: Oil and the End of Globalization*. Toronto: Random House.

———. 2009. *Why Your World is About to Get a Whole Lot Smaller: Oil and the End of Globalization*. Toronto: Random House.

———. 2012. *The Big Flatline: Oil and the No-Growth Economy*. Basingstoke: Palgrave Macmillan.

Rubin, Jeff, and Peter Buchanan. 2008. "What's the Real Cause of the Global Recession?" *StrategEcon*, October 31.

Rupert, Mark. 1995. *Producing Hegemony: The Politics of Mass Production and American Global Power*. Cambridge UK: Cambridge University Press.

Russi, Luigi. 2013. *Hungry Capital: The Financialization of Food*. Alresford: Zero Books.

Rutledge, Ian. 2005. *Addicted to Oil: America's Relentless Drive for Energy Security*. London: I. B Tauris.

Sawers, Larry. 1992. "The Navigation Acts Revisited." *Economic History Review* 45 (2): 262–84.

Sawvel, Franklin B. 1903. *The Complete Anas of Thomas Jefferson*. Cambridge, MA: Roundtable Press.

Schuster-Wallace, C. J., and R. Sandford. 2015. *Water in the World We Want*. United Nations University Institute for Water, Environment and Health and United Nations Office for Sustainable Development.

Scott, Bruce R. 2011. *Capitalism: Its Origins and Evolution as a System of Governance*. New York: Springer.

Shackleton, Michael. 1978. "Oil and the British Foreign Process." *Millennium: Journal of International Studies* 7 (2): 137–52.

Sheard, Paul. 2013. "Repeat after Me: Banks Cannot and Do Not Lend out Reserves." *Standard and Poor's Rating Service*: 1–15.

Sherwood, Marika. 2004. "Britain, the Slave Trade and Slavery, 1808–1843." *Race & Class* 46 (2): 54–77.

Shipley, Tyler. 2007. "Currency Wars: Oil, Iraq, and the Future of US Hegemony." *Studies in Political Economy* 79:7–33.

Shiva, Vandana. 2008. *Soil not Oil*. London: Zed Books.

Silber, Kenneth. 2009. "Stocks, Gold and War." In *Research*. New York: National Underwriter Company.

Silber, William L. 2008. *When Washington Shut Down Wall Street: The Great Financial Crisis of 1914 and the Origins of America's Monetary Supremacy*. Princeton, NJ: Princeton University Press.

SIPRI. 2013. "The SIPRI Top 100 Arms-Producing and Military Services Companies, 2013." Factsheet, December.

Smart, Allen, and Josephine Smart. 2003. "Urbanization and the Global Perspective." *Annual Review of Anthropology* 32:263–85.

Smart, Ian. 1977. "Oil, the Super-power and the Middle East." *International Affairs* 52 (1): 17–35.

Smil, Vaclav. 1994. *Energy in World History*. Boulder, CO: Westview Press.
———. 2006. *Energy*. Oxford: Oneworld Books.
———. 2011. "Energy: The Latest Infatuations." *American Scientist* 99 (3): 212–16.
Sobel, Robert. 1965. *A History of the New York Stock Market*. New York: The Free Press.
Sondhaus, Lawrence. 2001. *Naval Warfare 1815–1914*. London: Routledge.
Sosin, Jack M. 1964. "Imperial Regulation of Colonial Paper Money, 1764–1773." *Pennsylvania Magazine of History and Biography* 88 (2): 174–98.
Speth, Gus. 2009. *The Bridge at the End of the World: Capitalism, the Environment and Crossing from Crisis to Sustainability*. New Haven, CT: Yale University Press.
Spier, Fred. 2011. *Big History and the Future of Humanity*. Oxford: Wiley-Blackwell.
Spiro, David E. 1999. *The Hidden Hand of American Hegemony: Petrodollar Recycling and International Markets*. Ithaca, NY: Cornell University Press.
Stavrianos, L. S. 1981. *Global Rift: The Third World Comes of Age*. New York: William Morrow and Company, Inc.
Stohl, Rachel, and Suzette Grillot. 2009. *The International Arms Trade*. Cambridge: Polity Press.
Stokes, Doug. 2007. "Blood for Oil? Global Capital, Counter-insurgency and the Dual Logic of American Energy Security." *Review of International Studies* 33:245–64.
Stokes, Doug, and Sam Raphael. 2010. *Global Energy Security and American Hegemony*. Baltimore: Johns Hopkins University Press.
Strange, Susan. 1970. "International Economics and International Relations: A Case of Mutual Neglect." *International Affairs* 46 (2): 304–15.
———. 1988. *States and Markets*. London: Pinter Publishers.
Strauss, Sarah, Stephanie Rupp, and Thomas Love, eds. 2013. *Cultures of Energy: Power, Practices, Technologies*. Walnut Creek: Left Coast Press, Inc.
Sweezy, Paul. M. 1938. *Monopoly and Competition in the English Coal Trade 1550–1850*. Cambridge, MA: Harvard University Press.
———. 1954. *The Transition from Feudalism to Capitalism*. New York: Science and Society.
Sylla, Richard E. 1998. "U.S. Securities Markets and the Banking System." *Federal Reserve Bank of St. Louis Review* (May-June): 83–104.
Tarbell, Ida. 1904. *The History of the Standard Oil Company*. New York: McClure, Philips and Co.
Tate, William Edward. 1967. *The English Village Community and the Enclosure Movements*. London: Victor Gollancz.
Tawney, R. H. 1912. *The Agrarian Problem in the Sixteenth Century*. London: Longman, Green and Co.
———. 1926. *Religion and the Rise of Capitalism*. Harmondsworth: Penguin Books.
Tencer, Daniel. 2013. "Number of Cars Worldwide Surpasses 1 Billion: Can the World Handle This Many Wheels?" *Huffington Post*, February 19.
Therborn, Göran. 1977. "The Rule of Capital and the Rise of Democracy: Capital and Suffrage." *New Left Review* 1 (103): 3–41.

Thomas, Brinley. 1986. "Was There an Energy Crisis in Great Britain in the 17th Century?" *Explorations in Economic History* 23:124–52.

———. 2003. *The Industrial Revolution and the Atlantic Economy: Selected Essays.* London: Routledge.

Thomas, Hugh M. 2003a. *The English and the Normans.* Oxford: Oxford University Press.

———. 2003b. "The Significance and Fate of the Native English Landowners of 1086." *The English Historical Review* 118 (476): 303–33.

———. 2007. *The Norman Conquest: England after William the Conqueror.* Lanham, MD: Rowman & Littlefield Publishers, Inc.

Thompson, E. P. 1966. *The Making of the English Working Class.* New York: Vintage Books.

———. 1990. *Whigs and Hunters: The Origin of the Black Act.* London: Penguin.

Thompson, William. 2000. *The Emergence of the Global Political Economy.* London: Routledge.

Thorne, Susan. 2012. "Capitalism and Slavery Compensation." *Small Axe* 16 (1): 154–67.

Trainer, Ted. 2007. *Renewable Energy Cannot Sustain a Consumer Society.* New York: Springer.

Trebilcock, Clive. 1969. "'Spin-off' in British Economic History: Armaments and Industry, 1760–1914." *The Economic History Review* 22 (3): 474–90.

Trouillot, Michel-Rolph. 1995. *Silencing the Past: Power and the Production of History.* Boston: Beacon Press.

Turner, Louis. 1976. "The Oil Majors in World Politics." *International Affairs* 52 (3): 368–80.

Turner, Louis, and James Bedore. 1978. "Saudi Arabia: The Power of the Purse-Strings." *International Affairs* 54 (3): 405–20.

Turner, Raymond. 1921. "English Coal Industry in the Seventeenth and Eighteenth Centuries." *The American Historical Review* 27 (1): 1–23.

Tvedt, Terje. 2010. "Why England and Not China and India? Water Systems and the History of the Industrial Revolution." *Journal of Global History* 5:29–50.

Underhill, Geoffrey R. D. 2000. "State, Market, and Global Political Economy: Genealogy of an (Inter-?) Discipline." *International Affairs* 76 (4): 805–24.

UNDP. 2000. *World Energy Assessment: Energy and the Challenge of Sustainability.* New York: United Nations Development Programme.

UNEP and Bloomberg New Energy Finance. 2010. *Global Trends in Sustainable Energy Investment 2010: Analysis of Trends and Issues in Financing of Renewable Energy and Energy Efficiency.* http://bnef.com/free-publications/white-papers/.

Unger, R. W. 1984. "Energy Sources for the Dutch Golden Age: Peat, Wind and Coal." *Research in Economic History* 9:221–53.

Urry, John. 2012. "Do Mobile Lives Have a Future?" *Tijdschrift voor economische en sociale geografie* 103:566–76.

———. 2013. *Societies beyond Oil: Oil Dregs and Social Futures.* London: Zed Books.

US Senate. 1975. *Effect of Petrodollars on Financial Markets.* Hearing before the Subcommittee on Financial Markets of the Committee on Finance. United States Senate, 94th Congress. First Session, January.

van der Pilj, Kees. 1984. *The Making of an Atlantic Ruling Class*. London: Verso.
Veblen, Thorstein. 1904. *Theory of the Business Enterprise*. New York: Transaction Publishers.
———. 1923. *Absentee Ownership: Business Enterprise in Recent Times: The Case of America*. New York: Transaction Publishers.
Veit, Lawrence A. 1977. "Troubled World Economy." *Foreign Affairs* 55 (2): 263–79.
Victor, David G., David R. Hults, and Mark C. Thurber, eds. 2012. *Oil and Governance: State-Owned Enterprises and the World Energy Supply*. Cambridge: Cambridge University Press.
Visscher, Tommy L., and Jacob C. Seidell. 2001. "The Public Health Impact of Obesity." *Annual Review of Public Health* 22:355–75.
Vitalis, Robert. 2007. *America's Kingdom: Myth-making on the Saudi Oil Frontier*. Palo Alto, CA: Stanford University Press.
Vivoda, Vlado. 2009. "Resource Nationalism, Bargaining and International Oil Companies: Challenges and Change in the New Millennium." *New Political Economy* 14 (4): 517–34.
Vries, P. H. H. 2001. "Are Coal and Colonies Really Crucial? Kenneth Pomeranz and the Great Divergence." *Journal of World History* 12 (2): 407–46.
Wackernagel, Mathis, and William Rees. 1997. "Perceptual and Structural Barriers to Investing in Natural Capital: Economics from an Ecological Footprint Perspective." *Ecological Economics* 20 (3): 3–24.
Wallerstein, Immanuel. 1974. *The Modern World System I: Capitalist Agriculture and the Origins of the European World-Economy in the Sixteenth Century*. New York: Academic Press.
War on Want. 2008. *Banking on Bloodshed: UK High Street Banks' Complicity in the Arms Trade*. March. http://www.waronwant.org/attachments/Banking%20on%20Bloodshed.pdf.
Watts, Michael. 2012. "A Tale of Two Gulfs: Life, Death, and Dispossession along Two Oil Frontiers." *American Quarterly* 64 (3): 437–67.
Weaver, John C. 2003. *The Great Land Rush and the Making of the Modern World, 1650-1900*. Montreal: McGill-Queen's University Press.
Weisdorf, Jacob L. 2005. "From Foraging to Farming: Explaining the Neolithic Revolution." *Journal of Economic Surveys* 19 (4): 561–86.
Wennerlind, Carl. 2011. *Casualties of Credit: The English Financial Revolution, 1620-1720*. Cambridge, USA: Harvard University Press.
Werner, Richard. 2014. "Can Banks Individually Create Money out of Nothing? The Theories and Empirical Evidence." *International Review of Financial Analysis*, 36: 1–19.
White, Michael V. 1991. "A Biographical Puzzle: Why Did Jevons Write the Coal Question?" *Journal of the History of Economic Thought* 13 (2): 222–42.
———. 2004. "In the Lobby of the Energy Hotel: Jevons's Formulation of the Post-classical 'Economic Problem.'" *History of Political Economy* 36 (2): 227–71.
Wild, Martin. 2009. "Global Dimming and Brightening: A Review." *Journal of Geophysical Research* 114:1–31. doi:10.1029/2008JD011470.

Williams, Michael. 2006. *Deforesting the Earth: From Prehistory to Global Crisis*. Chicago: University of Chicago Press.

Williams, William Applemen. 2011. *The Contours of American History*. London: Verso.

Williamson, John, ed. 1990. *Latin American Adjustment: How Much Has Happened?* Washington, DC: Institute for International Economics.

Willrich, Mason. 1976. "Energy Independence for America." *International Affairs* 52 (1): 53–66.

Wilson, Charles. 1965. *England's Apprenticeship, 1603–1763*. Oxford: Oxford University Press.

Wolf, Martin. 2014. "We are Trapped in a Cycle of Credit Booms." *Financial Times*, October 7.

Wolff, E. N. 2010. "Recent Trends in Household Wealth in the United States: Rising Debt and the Middle-Class Squeeze—An Update to 2007." Working Paper No. 589. Annandale-on-Hudson, NY: The Levy Economics Institute of Bard College.

Woll, Peter, ed. 1996. *American Government: Readings and Cases*. New York: HarperCollins College Publishers.

Wood, Ellen Meiksins. 2002. *The Origin of Capitalism: A Longer View*. London: Verso.

———. 2012. *Liberty and Property: A Social History of Western Political Thought from Renaissance to Enlightenment*. London: Verso.

Wood, Ellen Meiksins, and Neal Wood. 1997. *A Trumpet of Sedition: Political Theory and the Rise of Capitalism, 1509–1688*. London: Pluto Press.

Wood, Gordon S. 1969. *The Creation of the American Republic: 1776–1787*. Chapel Hill: University of North Carolina Press.

World Bank. 2012. *Global Development Finance: External Debt of Developing Countries*. Washington, DC.

Wright, Robert E. 2008. *One Nation under Debt*. New York: McGraw Hill.

Wrigley, E. A. 2010. *Energy and the English Industrial Revolution*. Cambridge: Cambridge University Press.

Yergin, Daniel. 1991. *The Prize: The Epic Quest for Oil, Money and Power*. New York: Free Press.

———. 2006. "Ensuring Energy Security." *Foreign Affairs* 85 (2): 69–82.

Zalik, Anna. 2008. "Liquefied Natural Gas and Fossil Capitalism." *Monthly Review* 60 (6): 41–53.

Zehner, Ozzie. 2012. *Green Illusions: The Dirty Secrets of Clean Energy and the Future of Environmentalism*. Lincoln: University of Nebraska Press.

Žmolek, Michael Andrew. 2013. *Rethinking the Industrial Revolution*. Leiden: Brill.

Zook, Gorge Frederick. 1919. "The Company of Royal Adventurers of England Trading into Africa, 1660–1672." *The Journal of Negro History* 4 (2): 134–231.

Index

www.ingramcontent.com/pod-product-compliance
Ingram Content Group UK Ltd.
Pitfield, Milton Keynes, MK11 3LW, UK
UKHW040154091026
15587UKWH00018B/175

9 781783 480975